TO MOCK
A MOCKINGBIRD

And Other Logic Puzzles
Including an Amazing Adventure
in Combinatory Logic

RAYMOND M. SMULLYAN

With a Foreword by Melvin Fitting

WHAT

To the memory of HASKELL CURRY—
an early pioneer in combinatory logic
and an avid bird-watcher

CONTENTS

PART FOUR: SINGING BIRDS

PART FIVE: THE MASTER FOREST

PART SIX: THE GRAND QUESTION!

TO MOCK A MOCKINGBIRD

ACKNOWLEDGEMENTS

I wish to express my thanks to Nancy Spencer for expert typing and secretarial assistance, and to the Philosophy Department of Indiana University for providing me with ideal working conditions. My thanks go to Professor George Boolos at M.I.T. for reading this entire manuscript and for making many useful suggestions. Melvin Rosenthal, the production editor, has done a very conscientious job. My editor, Ann Close, has been wonderful, as usual, and enormously helpful with this whole project.

Raymond Smullyan

Elka Park, New York
November 1984

FORWARD TO MOCK A MOCKINGBIRD

Raymond Smullyan thought mathematics was one of the pleasures. Not always, all the time, in all ways. No pleasure is like that. But mathematics is often the producer of a sense of delight, of an unexpected insight, of a vision of deep beauty. One could see this in his face when he was teaching. Often there was a quiet smile of anticipation, knowing what was coming shortly for his class to appreciate. It saddened him to know that many people disliked mathematics, failed to see its attractions, even found it ugly. He believed that this was because many were scared of mathematics, and since mathematics was obviously a source of beauty and delight, there were so many wonderful things that people missed out on.

Puzzles are often much like mathematics. Indeed, it is hard to draw a clear line between the two. Raymond came to understand that while puzzles are, and must be, things of intrinsic interest, they could also be used to open new worlds of attractive mathematics to people, without their realizing it. Painlessly, so to speak.

Raymond applied this insight to a number of areas of mathematics, all having some connection with logic, but one book in particular (this one, if you haven't guessed) even led to results in the field itself. And this time Raymond did it all with birds.

Some kinds of mathematics require much background and development, broad knowledge of an area, before one can really begin. Calculus is a good example that an appreciable number of people are somewhat acquainted with. Before getting to the basics of calculus one must have a sufficient understanding of the real number system, the techniques of algebra, something of trigonometry, logarithms and exponentials. Then the ideas of calculus can begin to take their proper and impressive place. Most calculus courses, in fact, develop all of this more or less simultaneously. One must spend quite a bit of time with the entirety before one can begin to clearly see the truly astonishing power and simplicity that calculus has brought. This is not the setting for puzzles. It is the setting for exercises. Not the same thing at all.

You may never have heard about combinatory logic, but Raymond always had a fondness for it. As a source of puzzles, combinatory logic is a wonder. To get started, one can explain everything one needs to know in half a page! At that point one can already be asking really interesting questions. Of course if presented in a dry and mechanical way, it would come with the precondition of being boring. As you will see, Raymond replaced it with a discussion of a very peculiar forest (actually a succession of forests with varying properties) containing many kinds of birds. Each bird has a name, and responds with some name when it is called on, not necessarily its own. To quote from the book: "A certain enchanted forest is inhabited by talking birds. Given any birds A and B, if you call out the name of B to A, then A will respond by calling out the name of some bird to you; this bird we designate AB. Thus AB is the bird named by A upon hearing the name of B."

Some of these birds have special properties. For instance, the titular Mockingbird has the property that it imitates other birds in the following way. If you call the name of bird B to a mockingbird, it will respond just as bird B would have if you called its own name to it. In short, if M is a mockingbird then, for any bird B, MB is just BB. Roughly, a mockingbird mocks other birds by assuming they are egotists. It takes only a couple of kinds of birds, with simple but interesting properties, to produce a complex and fascinating forest.

But, where did all this come from? Why is it of any importance to mathematicians? Where did it go from here? Raymond is not going to tell you. He doesn't want to scare you away. Quite right too. So don't read this Forward beyond here, until you have already had fun with the book itself. Then think of the rest of this as an Afterward.

I am not going to give you a detailed history. That would be way too much for what is supposed to be a book of entertainment. Suffice it to say that combinatory logic began as a subject in the 1920s with the idea of simplifying certain aspects of formal logic, which had been introduced and was being investigated by that time. It was of interest only to specialists until computers came into being and the theoretical aspects of their behavior became important to understand. Indeed, the first proposed characterization of what, exactly, computers could and could not compute (in principle, of course) was in terms of an equivalent to combinatory logic. Since then, it is something that people studying computation are familiar with, not as the behavior of real machines, but as the study of an abstract and naturally defined class of idealized machines (idealized in a way similar to the way physics idealizes things by ignoring friction and similar annoyances). In a sense, in its creation combinatory logic provided the solution to problems yet to be formulated.

The previous paragraph, while brief, is enough to get a very general idea of the pre-history of Raymond's puzzles in this book. But the next question to ask is, "where did it go from here?" Surprisingly for a puzzle book intended to entertain, it did lead to consequences of a curious sort.

Automated theorem proving has a history going back to the early days of digital computers. Today it seems to be gradually getting enfolded into a very general and somewhat indistinct something called AI. But it's not at all clear that John McCarthy, who introduced the term "artificial intelligence" in the 1950s, would have seen present developments as having any direct relation to what he and his colleagues were doing. Be that as it may, by the mid 1980s significant theorem provers had been developed, and one in particular was ITP at Argonne National Laboratory. Smullyan's book, and especially one of its puzzles, was recognized "as an excellent test problem for automated theorem-proving programs." Smullyan's bird problems had a simple form that could be represented easily, and what background was needed was short and uncomplicated. I'll state the problem they chose, without sufficient explanation. You can find out everything you need in the book that follows. The problem is: given a bluebird and a warbler, find some sages. The characterizations of bluebirds, warblers, and sages are short, each occupying less than a line of type. The problem is easy to understand. It is a problem that is hard enough that Raymond left it as an open problem!

Rick Statman managed to show the existence of one such sage, and reported this to Raymond in a letter of 1986, then continued with further work based on Raymond's book, published in Statman (1989). Continuing with what Statman had started, but using their automated theorem prover, ITP, and a fair amount of tinkering from humans, 5 sages were discovered at Argonne, McCune and Wos (1987). The significance of this is not that the existence of a sage bird is an important fact, but that it provided a good testing ground for automated theorem proving. You could think of *To Mock a Mockingbird* as a good testing ground for yourself. Or, perhaps best of all, you could just have fun with it.

Melvin Fitting

March 28, 2023

REFERENCES

McCune, W., & Wos, L. (1987). A case study in automated theorem proving: finding Sages in combinatory logic. *Journal of Automated Reasoning*, 3, 91–107.

Statman, R. (1989). The word problem for Smullyan's Lark combinator is decidable. *Journal of Symbolic Computation*, 7, 103–112.

PREFACE

Before I tell you what this book is about, I would like to relate a true and delightful incident.

Shortly after the publication of my first puzzle book— called *What Is the Name of This Book?*—I received a letter from an unknown female suggesting an alternative solution to one of the puzzles, which I found more elegant than the one I had given. She closed the letter with "Love" and signed her name. I had absolutely no idea who she was, or whether she was married or single. I wrote back expressing my appreciation of her solution and asked whether I might use it in a subsequent edition. I also suggested that if she had not already graduated from college she might consider majoring in mathematics, since she showed definite mathematical talent. Shortly after, she replied: "Thank you for your gracious letter. You have my permission to use my solution. I am nine and a half years old and am in fifth grade."

She was particularly fond of puzzles about knights and knaves (truth-tellers and liars). Indeed, these puzzles have proved extremely popular with young and old alike, and I have accordingly devoted the first eight chapters of this book to new puzzles of this type. They range from extremely elementary to the very subtle metapuzzle in Chapter 8 on the Fountain of Youth. (Anyone solving *that* puzzle deserves to be knighted!) The rest of the book strikes out in a totally different direction, and goes into much deeper logical waters than any of my earlier puzzle books. You will learn some fascinating things about combinatory logic. This remarkable subject is currently playing an important role in computer science and artificial intelligence, and so this book is quite timely. (I didn't plan it that way; I just happened to be lucky!) Despite the profundity of the subject, it is no more difficult to learn than high school algebra or geometry. Since computer science has now found its way into the high school curriculum, could it be possible that combinatory logic will soon follow suit?

Combinatory logic is an abstract science dealing with objects called combinators. What their objects are need not be specified; the important thing is how they act upon each other. One is free to choose for one's "combinators" anything one likes (for example, computer programs). Well, I have chosen *birds* for my combinators—motivated, no doubt, by the memory of the late Professor Haskell Curry, who was both a great combinatory logician and an avid bird-watcher. The main reason I chose combinatory logic for the central theme of this book was not for its practical applications, of which there are many, but for its great entertainment value. Here is a field considered highly technical, yet perfectly available to the general public; it is chock-full of material from which one can cull excellent recreational puzzles, and at the same time it ties up with fundamental issues in modern logic. What could be better for a puzzle book?

PART ONE

LOGIC PUZZLES

CHAPTER 1

The Prize—and Other Puzzles

Three Little Puzzles

1 · The Flower Garden

In a certain flower garden, each flower was either red, yellow, or blue, and all three colors were represented. A statistician once visited the garden and made the observation that whatever three flowers you picked, at least one of them was bound to be red. A second statistician visited the garden and made the observation that whatever three flowers you picked, at least one was bound to be yellow.

Two logic students heard about this and got into an argument. The first student said: "It therefore follows that whatever three flowers you pick, at least one is bound to be blue, doesn't it?" The second student said: "Of course not!"

Which student was right, and why?

2 · What Question?

There is a question I could ask you that has a definite correct answer—either yes or no—but it is logically impossible for you to give the correct answer. You might *know* what the correct answer is, but you cannot give it. Anybody other than you might possibly be able to give the correct answer, but you cannot!

Can you figure out what question I could have in mind?

3 · Which Way Would You Bet?

Here is an old chestnut concerning probability: Choose your favorite baseball team and consider the scores it will make next season. Which do you bet will be the larger number—the *sum* of these scores or the *product* of these scores?

Speaking of probability and statistics, there is the story of a statistician who told a friend that he never took airplanes: "I have computed the probability that there will be a bomb on the plane," he explained, "and although this probability is low, it is still too high for my comfort." Two weeks later, the friend met the statistician on a plane. "How come you changed your theory?" he asked. "Oh, I didn't change my theory; it's just that I subsequently computed the probability that there would simultaneously be *two* bombs on a plane. This probability is low enough for my comfort. So now I simply carry my own bomb."

How Do You Win Your Prize?

4 · The Three Prizes

Suppose I offer to give you one of three prizes—Prize A, Prize B, or Prize C. Prize A is the best of the three, Prize B is middling, and Prize C is the *booby prize*. You are to make a statement; if the statement is true, then I promise to award you either Prize A or Prize B, but if your statement is false, then you get Prize C—the booby prize.

Of course it is easy for you to be sure to win either Prize A or Prize B; all you need say is: "Two plus two is four." But suppose you have your heart set on Prize A—what statement could you make which would force me to give you Prize A?

5 · A Fourth Prize Is Added

I now add a fourth prize—Prize D. This prize is also a booby prize. The conditions now are that if you make a true statement, I promise to give you either Prize A or Prize B, but if you make a false statement, you get one of the two booby prizes—Prize C or Prize D.

Suppose you happen to know in advance what the four prizes are, and for some reason or other, you like Prize C better than any of the other prizes.

Incidentally, this situation is not necessarily unrealistic. I recall that as a child I was at a birthday party where I won a prize, but became very envious of the kid who won the booby prize, because I liked his prize much more than mine! In fact, the booby prize seemed to be a general favorite, since we all spent most of the day playing with it.

Getting back to the present puzzle, what statement could you make that would force me to give you Prize C?

6 · You Wish to Confound Me!

Again, we have the four prizes of the last problem and the same conditions. Now, suppose you don't give a hoot for *any* of the prizes; you merely wish to confound me by making a statement that will force me to break my promise.

What statement would do this?

Note: This problem is essentially the same as the one known as the Sancho Panza paradox, a discussion of which is included in the solution.

SOLUTIONS

1 • The first student was right, and here is why. From the first statistician's report it follows that there cannot be more than one yellow flower, because if there were two yellows, you could pick two yellows and one blue, thus having a group of three flowers that contained no red. This is contrary to the report that every group of three is bound to contain at least one red flower. Therefore there cannot be more than one yellow flower. Similarly, there cannot be more than one blue flower, because if there were two blues, you could pick two blue flowers and one yellow and again have a group of three that contained no red. And so from the first statistician's report it follows that there is at most one yellow flower and one blue. And it follows from the report of the second statistician that there is at most one red flower, for if there were two reds, you could pick two reds and one blue, thus obtaining a group of three that contained no yellow. It also follows from the second report that there cannot be more than one blue, although we have already deduced this from the first report.

The upshot of all this is that there are only three flowers in the entire garden—one red, one yellow, and one blue! And so it is of course true that whatever three flowers you pick, one of them must be blue.

2 • Suppose I ask you: "Is no your answer to this question?" If you answer yes, then you are affirming that no is your answer to the question, which is of course wrong. If you answer no, then you are denying that no is your answer, although no was your answer. It is therefore impossible for you to answer the question correctly even though the question does have a correct answer: Either you answer no or you don't. If you do, then yes is the correct answer; if you don't, then no is the correct answer, but in neither case can you give the correct answer.

3 • The chances are that the sum will be the larger number, because your team will probably score at least one zero, and one zero makes the entire product zero.

4 • If you want to win Prize A, what you should say is: "I will not get Prize B." What can I do? If I give you Prize C, then your statement has turned out to be true—you didn't get Prize B—so I have given you the booby prize for making a true statement, which I cannot do. If I give you Prize B, then your statement has turned out to be false, but I can't give you Prize B for having made a false statement. Therefore I am forced to give you Prize A. You have then made a true statement—you didn't get Prize B—and have accordingly been awarded one of the two prizes offered for making a true statement.

Of course the statement "I will get either Prize A or Prize C" also works.

5 • To win Prize C, you need merely say: "I will get Prize D." I leave the proof to the reader.

6 • To force me to break my promise, you need merely say: "I will get one of the booby prizes." What can I do? If I give you a booby prize your statement has turned out to be true and I have broken my promise by giving you a booby prize. If I give you either Prize A or Prize B, then I have again violated the conditions of my promise, because you have made a false statement, and I should have given you a booby prize instead.

My promise was really a dangerous one, since I could have no way of knowing in advance whether or not I would be able to keep it. Whether or not I can keep it actually depends on what you do, as you have just seen. The situation is similar to the famous Sancho Panza paradox that Cervantes described in one episode of *Don Quixote.* In a certain town, the inhabitants had a decree that whenever a stranger crossed the bridge and entered the town, he was required to make a statement. If the statement was false, then the decree ordered that the stranger be hanged. If the statement was true, then he was to pass freely. What statement could the stranger make that would make it impossible for the decree to be carried out? The answer is for the stranger to say: "I will be hanged." It is then impossible for the inhabitants to carry out the decree.

CHAPTER 2

The Absentminded Logician

Only Three Words?

The puzzles of this chapter constitute as good an introduction to the logic of lying and truth-telling as I know. I will start with an old puzzle of mine and take off from there.

1 · John, James, and William

We are given three brothers named John, James, and William. John and James (the two J's) always lie, but William always tells the truth. The three are indistinguishable in appearance. You meet one of the three brothers on the street one day and wish to find out whether he is John (because John owes you money). You are allowed to ask him one question answerable by yes or no, but the question may not contain more than three words! What question would you ask?

2 · A Variant

Suppose we change the above conditions by making John and James both truthful and William a liar. Again you meet one of the three and wish to find out if he is John. Is there now a three-word yes/no question that can accomplish this?

3 · A More Subtle Puzzle

We now have only two brothers (identical twins). One of them is named Arthur and the other has a different name. One of the two always lies and the other always tells the truth, but we are not told whether Arthur is the liar or the truth-teller. One day you meet the two brothers together, and you wish to find out which one is Arthur. Note that you are *not* interested in finding out which one lies and which one tells the truth, but only in finding out which one is Arthur. You are allowed to ask just one of them a question answerable by yes or no, and again the question may not contain more than three words. What question would you ask?

· 4 ·

Suppose that instead of wanting to find out which one is Arthur, you want to find out whether Arthur is the liar or the truth-teller. Again there is a three-word question that will do this. What three-word question will work? There is a pretty symmetry between the solutions of this and the last problem!

· 5 ·

This time, all you are interested in finding out is which of the two brothers you meet is the liar and which is the truth-teller. You don't care which one is Arthur, or whether Arthur is the liar or the truth-teller. What three-word question will accomplish this?

· 6 ·

Next you are told to ask one of the brothers just one three-word question. If he answers yes, you will get a prize; if he answers no, then you get no prize. What question would you ask?

The Nelson Goodman Principle

If it were not for the restriction that the question contain no more than three words, all six of the above problems could be solved by a uniform method! This method is embodied in a famous puzzle invented about forty years ago by the philosopher Nelson Goodman. For those not familiar with the puzzle, here it is.

Given an individual who either always lies or always tells the truth, and given any proposition whose truth or falsity you wish to determine and whose truth or falsity is known by the individual, there is a way of determining this by asking just one yes/no question. For example, suppose the individual is standing at the fork of a road; one road leads to the town of Pleasantville, which you wish to visit, and the other road doesn't. The individual knows which road leads to Pleasantville, but he either always lies or always tells the truth. What question would you ask him to find out which is the correct road to Pleasantville?

Solution: If you ask him whether the left road is the correct one—i.e., the road that leads to Pleasantville—the question would be useless, since you have no idea whether he is a liar or a truth-teller. The right question to ask is: "Are you the type who could *claim* that the left road leads to Pleasantville?" After getting an answer, you will have no idea whether he is a liar or a truth-teller, but you *will* know which road to take! More specifically, if he answers yes, you should take the left road; if he answers no, you should take the right road. The proof of this is as follows:

Suppose he answers yes. Either he is truthful or he is lying. Suppose he is truthful. Then what he says is really so, hence he *is* the type who could claim that the left road leads to Pleasantville, and, since he is truthful, the left road really does lead there. On the other hand, if he is lying, then he is *not* the type who could claim that the left road leads to Pleasantville, since only one of the opposite type—a truth-teller—can make that claim. But since a truth-teller can make the claim, the claim must be correct, and so again the left road is the one leading to Pleasantville. This proves that regardless of whether the yes answer is the truth or a lie, the left road is the correct road to Pleasantville.

Suppose he answers no. If he is truthful, then he really is not the type who could claim that the left road leads to Pleasantville; only a liar would claim that it does. Since a liar would claim that it does, then it really doesn't, hence the right-hand road must lead to Pleasantville. On the other hand, if he is lying, then he *would* claim that the left road leads to Pleasantville, since he says he wouldn't, but since a liar would claim that the left road leads to Pleasantville, then it is really the right road that leads to Pleasantville. This proves that a no answer indicates that the right road leads to Pleasantville, regardless of whether the speaker is lying or telling the truth.

This puzzle somehow reminds me of an old Russian joke.

Boris and Vladimir are two old friends who meet unexpectedly one day on a train. The following conversation ensues:

Boris: Where are you going?

Vladimir: To Minsk.

Boris (indignantly): Why do you lie to me?

Vladimir: Why do you say I am lying?

Boris: You tell me you are going to Minsk in order to make me think you are going to Pinsk. But I know you are really going to Minsk!

Getting back to the Nelson Goodman principle, it is easy to see how we could uniformly solve the last six problems if we were not restricted to three-word questions. For example, in Problem 1, we could ask: "Are you the type who could claim that you are John?" The same goes for Problem 2. For Problem 3, we ask: "Are you the type who could claim you are Arthur?" For Problem 4, we could ask: "Are you the type who could claim that Arthur is truthful?"

In general, if you want to find out from a constant liar or constant truth-teller—you don't know which—whether a certain proposition p is true, you don't ask him: "Is p true?" Instead you ask him: "Are you the type who could *claim* that p is true?"

The Absentminded Logician

A certain logician, though absolutely brilliant in theoretical matters, was extremely unobservant and highly absent-minded. He met two beautiful identical-twin sisters named Teresa and Lenore. The two were indistinguishable in appearance, but Teresa always told the truth and Lenore always lied. The logician fell in love with one of them and married her, but unfortunately he forgot to find out her first name! The other sister didn't get married till a couple of years later. Quite shortly after the wedding, the logician had to go away for a logic conference. He returned a few days later. He then met one of the two sisters at a cocktail party and, of course, had no idea whether or not it was his wife. "I can find out in only one question," he thought proudly. "I'll simply use the Nelson Goodman principle and ask her if she is the type who could claim that she is my wife!" Then he had an even better idea: "I don't really have to be that elaborate and ask such a convoluted question. Why, I can find out if she is my wife by asking a much simpler question—in fact, one having only three words!"

· 7 ·

The logician was right! What three-word question answerable by yes or no should he ask to find out whether the lady he was addressing was his wife?

· 8 ·

A few days later the logician again met one of the two sisters at another cocktail party. He again didn't know whether it was his wife or his sister-in-law. "It's high time I find out once and for all my wife's first name," he thought. "I can ask this lady just one three-word yes/no question, and then I'll know!" What three-word question could he ask?

· 9 ·

Suppose that in the last problem, the logician had wanted to know both the identity of the lady he met and the first name of his wife. He is again restricted to asking only one question answerable by yes or no, but this time there is no restriction on the number of words in the question. Can you find a question that will work?

Epilogue: The logician was in fact married to the truthful sister, Teresa. Lenore's marriage, two years later, was a curious one: She detested her suitor, but when he one day asked her if she would like to be his wife, she, being a liar, had to answer yes. As a result, they got married!

The moral is that constant lying sometimes has its disadvantages.

SOLUTIONS

1 • The only three-word question I can think of that works is: "Are you James?" If you are addressing John, he will answer yes, since John lies, whereas both James and William would answer no—James because he lies, and William because he tells the truth. So a yes answer means that he is John and a no answer means that he is not John.

2 • The very same question—"Are you James?"—works, only a yes answer now indicates that he isn't John and a no answer indicates that he is John.

3 • A common wrong guess is: "Are you Arthur?" This question is quite useless here; the answer you get could be the truth or a lie, and you would still have no idea which one is really Arthur.

A question that works is: "Is Arthur truthful?" Arthur will surely answer yes to this question, because if Arthur is truthful, he will truthfully claim that Arthur is truthful, and if Arthur is not truthful, then he will falsely claim that Arthur is truthful. So regardless of whether Arthur is truthful or whether he lies, he will certainly *claim* that Arthur is truthful. On the other hand, Arthur's brother—call him Henry—will claim that Arthur is not truthful, because if Henry is truthful, then Arthur is really not truthful and Henry will truthfully claim that Arthur is not. And if Henry lies, then Arthur really is truthful, in which case Henry will falsely claim that Arthur is *not* truthful. So whether Henry is truthful or not, he will surely claim that Arthur is not truthful. In summary, Arthur will claim that Arthur is truthful and Arthur's brother will claim that Arthur is not truthful. So if you ask one of the brothers whether Arthur is truthful, and if you get yes for an answer, you will know that you are speaking to Arthur; if you get no for an answer, you will know that you are speaking to Arthur's brother.

Incidentally, there is another three-word question that works: "Does Arthur lie?" A yes answer to that question would indicate that you are *not* speaking to Arthur, and a no answer would indicate that you *are* speaking to Arthur. I leave the verification of this to the reader.

4 • To find out whether Arthur is truthful, all you need to ask is: "Are you Arthur?" Suppose you get the answer yes. If it is a truthful answer, then the one addressed really is Arthur, in which case Arthur is the truthful brother. If the answer is a lie, then the answerer is not really Arthur, in which case Arthur must be the other one, again the truthful brother. So regardless of whether the answer is truthful or a lie, a yes answer indicates that Arthur—whichever one he is—must be truthful. What if you get no for an answer? Well, if it is a truthful answer, then the speaker is not Arthur, but since he is truthful, Arthur must be the brother who lies. On the other hand, if the no answer was a lie, then the speaker really is Arthur, in which case Arthur just told a lie. So a no answer, whether it is the truth or a lie, indicates that Arthur is the liar.

5 • Just ask him: "Do you exist?"

6 • Just ask: "Are you truthful?" Both constant truth-tellers and constant liars will answer yes to that question.

7 • We recall that his wife's sister was not married at the time. A three-word question that works is: "Is Teresa married?" Suppose the lady answers yes. She is either Teresa or Lenore. Suppose she is Teresa. Then the answer is truthful, hence Teresa is really married, and the lady addressed is married and his wife. If she is Lenore, the answer is a lie; Teresa is not really married, so Lenore—who is the lady addressed—is married, hence again the lady addressed is his wife. So a yes answer indicates that he is speaking to his wife, regardless of whether the answer is the truth or a lie. I leave it to the reader to verify that a no answer indicates that he is speaking to his wife's sister.

8 • The question to ask now is: "Are *you* married?" Suppose she answers yes. Again, she is either Teresa or Lenore. Suppose she is Teresa. Then the answer is truthful, hence the lady addressed is married, and since she is Teresa, he is married to Teresa. But what if the lady addressed is Lenore? Then the answer is a lie, hence the lady addressed is not really married, and he is married to the other lady, again Teresa. So in either case, a yes answer indicates that his wife's name is Teresa.

I again leave it to the reader to verify that a no answer indicates that his wife's name is Lenore.

9 • No, because no such question exists!

You see, in each of the preceding problems, we were trying to find out which of two possibilities holds, but in this problem, we are trying to find out which of *four* possibilities holds. (The four possibilities are that the lady addressed is Teresa, his wife; that she is Lenore, his wife; that she is Teresa, his sister-in-law; and that she is Lenore, his sister-in-law.) However, a yes/no question can elicit only two possible responses, and with only two possible responses it is impossible to determine which of *four* possibilities holds.

CHAPTER 3

The Barber of Seville

1 · A Double Barber Paradox

Some of you are familiar with Bertrand Russell's paradox about a barber of a certain town who shaved all of and only those inhabitants who did not shave themselves. In other words, given any inhabitant X who didn't shave himself, the barber was certain to shave him. But the barber *never* shaved any inhabitant X who shaved himself. The problem is: Did the barber shave himself or didn't he?

If the barber shaved himself, then he violated his rule by shaving someone—namely himself—who shaved himself. This is impossible; hence he didn't shave himself. But if he didn't shave himself, then he failed to shave someone—again himself—who didn't shave himself. So it is also impossible that he didn't shave himself. Therefore, did he shave himself or didn't he?

The solution is simply that there couldn't be any such barber. The assumption that such a barber exists leads to a contradiction; hence it is false.

The barber paradox bears a close relation to the famous liar paradox. Consider the sentence written in the following box:

> The Sentence in This Box Is False.

Is the sentence true or false? If it is true, then what it asserts is really the case, which means that it must be false, because that is what it asserts. On the other hand, if it is false, then what it asserts is not the case, which means that it is not the case that it is false; hence it must be true. So the assumption that it is false again leads to a contradiction.

There is also Phillip Jourdain's "double" version of the liar paradox. Consider a card with the following sentence written on one side:

Side 1

The Sentence on the Other Side Is False.

When the card is turned over, it reads:

Side 2

The Sentence on the Other Side Is True.

Side 2, by asserting that the sentence on Side 1 is true, is asserting that what Side 1 says is really the case—in other words, that the sentence on Side 2 is false. So the sentence on Side 2 is indirectly asserting its own falsity, and we have the same paradoxical situation.

I have thought of a paradox that bears much the same relation to Russell's barber paradox as Jourdain's double liar paradox bears to the older liar paradox. You should read all three parts of the problem before turning to the solution.

Suppose I tell you that a certain town contains a barber named Arturo and that given any inhabitant X *other than Arturo himself,* Arturo shaves X if and only if X doesn't shave Arturo. In other words, if Arturo shaves X, then X doesn't shave Arturo, but if Arturo doesn't shave X, then X does shave Arturo. Does this lead to a paradox?

Suppose, instead, I had told you that the town contains a barber named Roberto and that for any inhabitant X, Roberto shaves X if and only if X *does* shave Roberto. In other words,

if Roberto shaves X, then X shaves Roberto, and if X shaves Roberto, then Roberto shaves X. Does this lead to a paradox? Now, suppose I told you that the town contains both barbers—Arturo *and* Roberto—satisfying the above conditions. Does this lead to a paradox? Why or why not?

2 · WHAT ABOUT THIS ONE?

Suppose I told you that the town contains two barbers, Arturo and Roberto, and that Arturo shaves all and only those inhabitants who shave Roberto, and Roberto shaves all and only those inhabitants who don't shave Arturo. Does this lead to a paradox?

3 · BARBER FOR A DAY

A certain town contained exactly 365 male inhabitants. During one year, which was not a leap year, it was agreed that on each day one man would be official barber for the day. No man served as official barber for more than one day. Also, the official barber on a given day was not necessarily the *only* person who shaved people on that day; nonbarbers could also do some shaving.

Now, it is given that on any day, the official barber for that day—call him X—shaved at least one person. Let X* be the *first* person shaved by X on the day when X was official barber. We are also given that for any day D, there is a day E such that for any male inhabitants X and Y, if X shaved Y on day E, then X* shaved Y on day D.

Now, the above conditions certainly lead to no paradox, but they do lead to an interesting conclusion, namely that on each day at least one person shaved himself. How do you prove this?

4 · THE BARBERS' CLUB

There is a certain club called the Barbers' Club. The following facts are known about it.

Fact 1: Every member of the club has shaved at least one member.

Fact 2: No member has ever shaved himself.

Fact 3: No member has ever been shaved by more than one member.

Fact 4: There is one member who has never been shaved at all.

The number of members of this club has been kept a strict secret. One rumor has it that there are less than a thousand members. Another rumor has it that there are more than a thousand members. Which of the two rumors is correct?

5 · ANOTHER BARBERS' CLUB

Here is a well-known logic puzzle, appropriately dressed for the occasion.

Another barbers' club obeys the following conditions:

Condition 1: If any member has shaved any member—whether himself or another—then all the members have shaved him, though not necessarily all at the same time.

Condition 2: Four of the members are named Guido, Lorenzo, Petruchio, and Cesare.

Condition 3: Guido has shaved Cesare.

Has Petruchio shaved Lorenzo or not?

6 · THE EXCLUSIVE CLUB

There is another club known as the Exclusive Club. A person is a member of this club if and only if he doesn't shave anyone who shaves him.

A certain barber named Cardano once boasted that he had shaved every member of the Exclusive Club and no one else. Prove that his boast involves a logical impossibility.

7 · THE BARBER OF SEVILLE

Any resemblance between the Seville of this story and the famous Seville of Spain (which in fact there isn't) is purely coincidental.

In this mythical town of Seville, the male inhabitants wear wigs on those and only those days when they feel like it. No two inhabitants behave alike on *all* days; that is, given any two male inhabitants, there is at least one day on which one of them wears a wig and the other doesn't.

Given any male inhabitants X and Y, inhabitant Y is said to be a *follower* of X if Y wears a wig on all days that X does. Also, given any inhabitants X, Y, and Z, inhabitant Z is said to be a follower of X and Y if Z wears a wig on all days that X and Y both do.

Five of the inhabitants are named Alfredo, Bernardo, Benito, Roberto, and Ramano. The following facts are known about them:

Fact 1: Bernardo and Benito are opposite in their wig-wearing habits; that is, on any given day, one of them wears a wig and the other one doesn't.

Fact 2: Roberto and Ramano are likewise opposites.

Fact 3: Ramano wears a wig on those and only those days when Alfredo and Benito both wear one.

Seville has exactly one barber, and the following facts are known about him:

Fact 4: Bernardo is a follower of Alfredo and the barber.

Fact 5: Given any male inhabitant X, if Bernardo is a follower of Alfredo and X, then the barber is a follower of X alone.

Alfredo wears only black wigs; Bernardo wears only white wigs; Benito wears only gray wigs; Roberto wears only red wigs; and Ramano wears only brown wigs.

One Easter morning, the barber was seen wearing a wig. What color was he wearing?

Solutions

1 • The existence of Arturo alone creates no paradox, nor does the existence of Roberto alone. But it is impossible that they can both exist in the same town. (They are what Ambrose Bierce would call "incompossible"; see note following solution.) Here is the reason why:

For any person X other than Arturo, Arturo shaves X if and only if X doesn't shave Arturo. This is true for *any* person X other than Arturo; so, in particular, it is true if X is Roberto. Then, taking Roberto for X, we have:

1. Arturo shaves Roberto if and only if Roberto doesn't shave Arturo. In other words, if Arturo shaves Roberto, then Roberto does not shave

him back, but if Arturo doesn't shave Roberto, then Roberto does shave Arturo. Stated in still different terms, one of them shaves the other, but the other does not shave him back.

On the other hand, we are given that for any person X, Roberto shaves X if and only if X *does* shave Roberto. Taking Arturo for X, we have

2. Roberto shaves Arturo if and only if Arturo *does* shave Roberto.

Statement 2 says that Arturo and Roberto either both shave each other or neither one shaves the other. This is the very opposite of Statement 1, which says that one of them shaves the other, but the other doesn't shave him back. Therefore, the given conditions are impossible; Arturo and Roberto are really "incompossible."

Note: In *The Devil's Dictionary,* Ambrose Bierce gives the following definition:

> INCOMPOSSIBLE, adj. Unable to exist if something else exists. Two things are incompossible when the world of being has scope enough for one of them but not enough for both—as Walt Whitman's poetry and God's mercy to man. Incompossibility, it will be seen, is only incompatibility let loose. Instead of such low language as "Go heel yourself; I mean to kill you on sight," the words, "Sir, we are incompossible," would convey an equally significant intimation and in stately courtesy are altogether superior.

2 • No, this is no paradox. It could be that Roberto shaves himself, Arturo shaves Roberto, Arturo doesn't shave himself, and Roberto doesn't shave Arturo. The other X's in the town don't really matter; indeed, Arturo and Roberto could just as well be the town's only inhabitants.

3 • Take any day D. We are given that E is a day such that for any inhabitants X and Y, if X shaved Y on day E, then X* shaved Y on day D. Now, let X be the official barber on day E. This means X shaved X* on day E (in fact, X* was the first one to be shaved on day E). Then, taking X* for Y, if X shaved X* on day E, then X* shaved X* on day D. And X did shave X^* on day E. Therefore, X^* shaved X^* on day D; in other words, on day D, X^* shaved himself.

The upshot is that for any day D, we let day E be as given in the conditions of the problem. Then it was not the barber of day E who necessarily shaved himself on day D, but the first one shaved by the barber on day E who must have shaved himself on day D.

4 • It is the second rumor—that there are more than a thousand members—which must be correct. Indeed, there are a lot more; the only way the given conditions can hold is that there be an *infinite* number of members! Let us see why:

By Fact 4, we know there is a member—call him B_1—who has never been shaved at all. Now, B_1 has shaved at least one member, but this member could not be B_1, since B_1 has never been shaved, so it must be someone else, whom we will call B_2. Now, B_2 has shaved someone, but it couldn't be B_1, who has never been shaved, or B_2, since no one shaved himself; so it must be a new person—B_3. Now, B_3 has also shaved someone, but it was not B_1, who has never been shaved, nor B_2, since B_2 was shaved by Bi, nor himself, B_3. So it was a new person—B_4. Again, B_4 couldn't have shaved B_1, nor B_2, who was shaved by Bi, nor B_3, who was shaved by B_2, nor B_4, himself, so it was another person, B_5. Applying the same kind of argument to B_5, we see that he must have shaved some person B_6 different from each of B_1, B_2, B_3, B_4, B_5. Then B_6 shaved some new person, B_7, and so on. In this way we generate an infinite sequence of distinct members, so no *finite* number can suffice.

5 • Since Guido has shaved Cesare, then Guido has shaved at least one member. Therefore, all the members have shaved Guido. In particular, Lorenzo has shaved Guido. Therefore, Lorenzo has shaved at least one member, so all the members have shaved Lorenzo. In particular, Petruchio has shaved Lorenzo.

What really follows from the three conditions is that every member of the club has shaved every member!

6 • Suppose Cardano's claim is true—we get the following contradiction: To begin with, no member of the Exclusive Club has ever shaved himself, because he has never shaved anyone—including himself—who has shaved him. Now, suppose Cardano is a member of the club. Then he has not shaved himself, as we have just proved, hence he has failed to shave at least one member of the club—namely himself. This is contrary to his claim that he has shaved every member of the club. Therefore Cardano cannot be a member of the club.

Since Cardano is not a member of the club, then he has shaved at least one person who has shaved him. Let Antonio be such a person. Then Antonio, a club member, has shaved someone—namely Cardano—who has shaved him, which no club member can do! This is clearly a contradiction, hence Cardano's story doesn't hold water.

7 • *Step 1:* First, we prove that Roberto is a follower of the barber.

Well, consider any day on which the barber wears a wig. Either Alfredo wears a wig on that day or he doesn't. Suppose Alfredo does. Then Bernardo also wears a wig on that day, because Bernardo is a follower of Alfredo and the barber. So Benito can't wear a wig on that day, because he is opposite to Bernardo. Then Ramano can't wear a wig on that day, because he wears wigs only on those days when Alfredo and Benito both do, and Benito doesn't have one on this day. Since Ramano doesn't wear a wig on this day, then Roberto must, because Roberto is opposite to Ramano. This proves that on any day on which the barber wears a wig, if Alfredo also does, then so does Roberto.

Now, what about a day on which the barber wears a wig but Alfredo doesn't? Well, since Alfredo doesn't, then it certainly is not the case that Alfredo and Benito both do; hence Ramano doesn't, by Fact 3, and therefore Roberto does, by Fact 2. So Roberto wears a wig on any day that the barber does and Alfredo doesn't—indeed, he wears a wig on all days that Alfredo doesn't, regardless of the barber.

This proves that on any day on which the barber wears a wig, Roberto also does, regardless of whether Alfredo does or does not wear a wig on that day. So Roberto is indeed a follower of the barber.

Step 2: We next prove that Bernardo is a follower of Alfredo and Roberto.

Consider any day on which Alfredo and Roberto both wear wigs. Benito can't wear a wig on that day, because if he did, then Alfredo and Benito both would have them on, which by Fact 3 would mean that Ramano also does, and we would have Ramano wearing a wig on the same day as Roberto, which is contrary to Fact 2. Therefore, on any day on which Alfredo and Roberto both wear wigs, Benito doesn't, and hence Bernardo does, according to Fact 1. This proves that Bernardo is a follower of Alfredo and Roberto.

Step 3: Now we are in a position to show that the barber is a follower of Roberto, which is the converse of what we proved in Step 1.

To do this, we use Fact 5 for the first time. Fact 5 is true for *every* male inhabitant X, so in particular it holds if X is Roberto. Therefore, we know that if Bernardo is a follower of Alfredo and Roberto, then the barber is a follower of Roberto alone. And Bernardo *is* a follower of Alfredo and Roberto, by Step 2. Therefore, the barber is a follower of Roberto.

Step 4: Now we know that Roberto is a follower of the barber, by Step 1, and that the barber is also a follower of Roberto, by Step 3. Therefore, Roberto and the barber wear wigs on exactly the same day. But we were given that no two different people wear wigs on exactly the same days. Hence Roberto and the barber must be the same person! Finally, since Roberto wears only red wigs, the barber can wear only red wigs. So the answer to the problem is *red*.

CHAPTER 4

THE MYSTERY OF THE PHOTOGRAPH

INTRODUCTION

For the benefit of my new readers, I had best review an old puzzle of mine.

Suppose we consider, as in Chapter 2, a pair of brothers, one of whom always tells the truth and the other of whom always lies. But now we have a new complication. The truth-teller is completely accurate in all his judgments; all true propositions he knows to be true, and all false propositions he knows to be false. On the other hand, the lying brother is totally inaccurate in his judgments; all true propositions he believes to be false and all false propositions he believes to be true. Now, in fact, whatever question you ask of either brother, you will get the same answer. For example, suppose you ask whether two plus two equals four. The accurate truth-teller will know that two plus two equals four and will honestly answer yes. The inaccurate liar will erroneously believe that two plus two *doesn't* equal four, but then he will lie and claim that it *does* equal four, and so he will also answer yes.

The situation is reminiscent of an incident 1 once read about in a psychiatric journal and reported in *What Is the Name of This Book?*

The doctors of a certain mental institution were thinking of releasing a schizophrenic patient. They decided first to give him a lie detector test. One of the two questions they asked him was: "Are you Napoleon?" He replied, "No." The machine showed he was lying!

Getting back to my puzzle, suppose the two brothers are identical twins who are indistinguishable in appearance. You meet one of them alone and wish to find out whether he is the accurate truth-teller or the inaccurate liar. Is it possible to do this by asking any number of yes/no questions? According to one argument, it is not possible, because whatever question you ask of either brother, you will get the same answer (as I have proved). But there is another argument—which I won't yet tell you—that it *is* possible. Is it really possible or isn't it?

The answer is yes, it is possible; moreover, it can be done in only one question! All you need ask is: "Are you the accurate truth-teller?" If he is, then he will know that he is, since he is accurate, and will honestly reply yes. But if he is the inaccurate liar, then he will *believe* he's the accurate truth-teller, since his beliefs are all wrong, and will lie about his beliefs and answer no. So if he answers yes, you will know he is the accurate truth-teller; if he answers no, you will know he is the inaccurate liar.

But doesn't this raise a paradox? On the one hand, I have proved that the two brothers will give the same answer to the same question, yet I have just exhibited a question to which they would give different answers! How can this be? Was my first proof fallacious?

The answer is no, the proof was perfectly valid; both brothers will indeed respond the same way to the same question, and in fact will give the correct answer. The whole point is that the six words "Are you the accurate truth-teller?" when asked of one person constitute a *different* question than when they are asked of another, because they contain the variable term "you," whose denotation depends on the person addressed! And so you are not really asking the same question, even though the sequence of words is the same.

The technical adjective for words like "you," "I," "this," "that," and "now" is *indexical*. Their denotations are not absolute, but depend upon their contexts. This "indexical" principle was delightfully used by Ambrose Bierce in his book *The Devil's Dictionary*. After defining the word "I," he goes on to say: "The plural of I is said to be *we*, but how there can be more than one myself is doubtless clearer to the grammarians than it is to the author of this incomparable dictionary."

THE CASE OF THE FOUR BROTHERS

Now we have *four* brothers, named Arthur, Bernard, Charles, and David. The four are quadruplets and are indistinguishable in appearance. Arthur is an accurate truth-teller; Bernard is an inaccurate truth-teller (he is totally deluded in all his beliefs but always states honestly what he does believe); Charles is an accurate liar (all his beliefs are correct, but he lies about every one of them); and David is an inaccurate liar (he is both deluded and dishonest; he tries to give you false information but he is unable to!).

You see, of course, that Arthur and David will both give correct answers to any question asked, whereas both Bernard and Charles will give the wrong answer to any question asked.

1 · A SIMPLE STARTER

Suppose you meet one of the four brothers one day on the street. You wish to find out his first name, and you are allowed to ask only yes/no questions. What is the smallest number of questions you need ask, and what would the questions be?

· 2 ·

Arthur and Bernard are both married; the other two brothers are not. Arthur and Charles are both wealthy; the other two brothers are not.

You meet one of the four brothers one day and wish to find out whether he is married. What yes/no question would you ask? This can be done with a three-word question!

· 3 ·

Suppose, instead, you wanted to find out if he is wealthy. What question would you ask?

· 4 ·

I once met one of the four brothers and asked him a yes/no question. I should have realized before I asked it that the question was pointless, because I could have known in advance what the answer would be. Can you supply such a question?

· 5 ·

Suppose someone makes you the following offer. You are to interview one of the four brothers and try to find out which one he is. You may ask him one question, or you may ask him two questions, but you must decide in advance which you will do. If you choose the two-question option and ascertain his identity, you will be given a prize of a hundred dollars. But if you choose the one-question option and can determine his identity, you will get a thousand dollars. However, under this option, if you fail after the first question, you are not allowed to ask a second.

From the point of view of pure mathematical probability, would you choose the one-question option or the two-question option?

· 6 ·

In preparation for the two special puzzles, 8 and 9, that follow, I wish to illustrate a basic principle.

As you already know, if you ask any of the four brothers whether two plus two equals four, Arthur and David will answer yes, and Bernard and Charles will answer no. Now, suppose you instead ask: "Do you believe that two plus two equals four?" What will each brother answer?

· 7 ·

Suppose you ask one of the brothers whether two plus two equals four and he answers no. Then you ask him whether he *believes* that two plus two equals four and he answers yes. Which of the four brothers is it?

Two Special Puzzles

8 · A Metapuzzle

One day a logician came across one of the four brothers and asked him, "Who are you?" The brother identified himself as either Arthur, Bernard, Charles, or David, and the logician then knew who he was.

A few minutes later, the same brother was met by a second logician, who asked him, "Who do you believe you are?" The brother answered, again either Arthur, Bernard, Charles, or David, and the second logician then knew who he was.

Who was he?

9 · The Mystery of the Photograph

If you ever visit these four brothers at their home, you will notice a photograph of one of them in the living room. If you ask each of them whether it is *his* photograph, three of them will answer no and one will answer yes. If you ask each one whether he *believes* it is his photograph, then again three will answer no and one will answer yes. Whose photograph is it?

Solutions

1 • Two questions are enough, and these can be chosen in many ways. Here is one sequence:

You first ask the brother you meet whether two plus two equals four. If he answers yes, then you know that he answers all questions correctly and must therefore be Arthur or David. Then you simply ask him whether he is Arthur and you abide by what he answers. If he answers no to your first question, then you know that he answers all questions incorrectly and that he must be either Bernard or Charles. Then you ask him if he is Bernard and you abide by the opposite of what he answers.

2 • To find out whether he is married, you need only ask him: "Are you wealthy?" Arthur will answer yes, since he is wealthy and gives correct answers; Bernard will also answer yes, since he is not wealthy and gives wrong answers; Charles will answer no, since he is wealthy but gives wrong answers; and David will answer no, since he is not wealthy and gives correct answers. And so a married brother will answer yes and an unmarried brother will answer no.

3 • To find out if he is wealthy, you ask: "Are you married?" As the reader can check, Arthur and Charles, the wealthy brothers, will answer yes and Bernard and David will answer no.

4 • The question I stupidly asked was: "Are you either Arthur or David?" I should have known that any of the four brothers would answer yes, since if he was either Arthur or David, he would correctly answer yes; if he was Bernard or Charles, he would incorrectly answer yes.

5 • You are better off choosing the one-question option! With the two-question option, you have a certainty of winning a hundred dollars by following the procedure of the solution of Problem 1, but with the one-question option, you have a one-out-of-four chance to win a thousand dollars. Just ask the in terviewee if he is Bernard. If he is Charles, he will answer yes; each of the other three will answer no, as you can check. So if you get yes for an answer, you will know he is Charles, and you can claim your prize. If you get no for an answer, you won't know which of the other three he is, and so you cannot then collect. But a one-quarter chance at a thousand dollars is mathematically better odds than a certain win of a hundred dollars.

6 • Remember that you are asking the brother whether he *believes* that two plus two equals four. Arthur would obviously answer yes. Now, what about Bernard? Many readers will be tempted to say that Bernard will answer no, but this is not so! Bernard would also answer yes, and here is why:

Bernard, being inaccurate, doesn't believe that two plus two equals four. But Bernard is inaccurate in *all* his beliefs—even his beliefs about his own beliefs! And so he will wrongly believe that he *does* believe that two plus two equals four! Then, being a truth-teller, he will honestly claim what he believes—namely that he *believes* that two plus two equals four.

A possibly simpler way of seeing the situation is this: Bernard doesn't believe that two plus two equals four, and so the *correct* answer to the question is no. But he gives only incorrect answers to questions, so he answers yes.

Now, with Charles, it is a different story. Charles *does* believe that two plus two equals four, and being accurate, he believes that he believes that two plus two equals four. But he is a liar, hence he denies that he believes that two plus two equals four, and answers no.

David doesn't believe that two plus two equals four, and he answers all questions correctly, so he will answer no.

In summary, Arthur and Bernard, the truth-tellers, answer yes and Charles and David, the liars, answer no to the question of whether they *believe* that two plus two is equal to four.

7 • Only Bernard and Charles can deny that two plus two equals four. Only Arthur and Bernard can claim to *believe* that two plus two equals four, as we have seen from the last problem. So the only brother who could answer no to the first question and yes to the second is Bernard.

8 • We leave it to the reader to verify the following four facts:

1. Arthur, Charles, and Bernard could each claim to be Arthur.
2. Only Charles could claim to be Bernard.
3. Only Bernard could claim to be Charles.
4. Bernard, Charles, and David could each claim to be David.

Therefore, if the first logician had received either the answer "Arthur" or "David," he couldn't have known which brother he was addressing. But he did know, so he either got the answer "Bernard" and knew that the brother was really Charles, or he got the answer "Charles" and knew that the speaker was really Bernard. So now *you* know that the brother was either Bernard or Charles, but you don't know which, although the first logician did.

Now for the question of the second logician. Using principles illustrated in the last two problems, the reader should be able to verify the following four facts:

1. Arthur, Charles, and David could each claim to *believe* he is Arthur.

2. Bernard, Charles, and David could each claim to believe that he is Bernard.

3. Only David could claim to believe he is Charles.

4. Only Charles could claim to believe he is David.

So the only way the second logician could have been sure of the identity of the brother is that either he got the answer "Charles" and knew that the speaker was David, or he got the answer "David" and knew that the speaker was Charles. We already know that he is not David—because he is either Bernard or Charles—so he must be Charles.

9 • As the reader can verify, if the photograph is of Arthur, then three of the brothers, Arthur, Bernard, and Charles, would answer yes to the first question, and therefore the photograph is not of Arthur. If the photograph were of David, again you would get three yes answers to the first question, from Bernard, Charles, and David, so the photograph is not of David. With Bernard's photograph, you would get three no answers, by Arthur, Bernard, and David, and with Charles's photograph you would also get three no answers, by Arthur, Charles, and David. And so from the fact that three of the answers to the first question are no, the photograph must be of Bernard or Charles.

As to the second question, if the photograph is of Bernard, you would get only one no answer, from Arthur, but with Charles's photograph, you would get three no answers, from Arthur, Bernard, and Charles. Therefore the photograph is of Charles.

PART TWO

KNIGHTS, KNAVES, AND THE FOUNTAIN OF YOUTH

CHAPTER 5

SOME UNUSUAL KNIGHTS AND KNAVES

My earlier puzzle books—*What Is the Name of This Book?*, *The Lady or the Tiger?*, and *Alice in Puzzle-Land*—are chock-full of puzzles about an island in which every inhabitant is either a knight or a knave, and knights make only true statements and knaves only false ones. These puzzles have proved popular, and I give some new ones in this chapter. First, however, we will consider five questions that will serve both as an introduction to knight-knave logic for those not familiar with it and as a brief refresher course for those who are. Answers are given following the fifth question.

Question 1: Is it possible for any inhabitant of this island to claim that he is a knave?

Question 2: Is it possible for an inhabitant of the island to claim that he and his brother are both knaves?

Question 3: Suppose an inhabitant A says about himself and his brother B: "At least one of us is a knave." What type is A and what type is B?

Question 4: Suppose A instead says: "Exactly one of us is a knave." What can be deduced about A and what can be deduced about B?

Question 5: Suppose A instead says: "My brother and I are the same type; we are either both knights or both knaves." What could then be deduced about A and B? Suppose A had instead said: "My brother and I are different types." What can then be deduced?

Answer 1: No; no inhabitant can claim to be a knave because no knight would lie and say he is a knave and no knave would truthfully admit to being a knave.

Answer 2: This question has provoked a good deal of controversy! Some claim that anyone who says that he and his brother are both knaves is certainly claiming that he is a knave, which is not possible, as we have seen in the answer to Question 1. Therefore, they conclude, no inhabitant can claim that he and his brother are both knaves.

This argument is wrong! Suppose an inhabitant A is a knave and his brother B is a knight. Then it is *false* that he and his brother are both knaves, hence he, as a knave, is certainly capable of making that false statement. Therefore it *is* possible for an inhabitant to claim that he and his brother are both knaves, but only if he is a knave and his brother is a knight.

This illustrates a curious principle about the logic of lying and truth-telling: Normally, if a truthful person claims that both of two statements are true, then he will certainly claim that each of the statements is true separately. But with a constant liar, the matter is different. Consider the following two statements: (1) My brother is a knave; (2) I am a knave. A knave could claim that (1) and (2) together are *both* true, provided his brother is actually a knight, but he cannot claim (1) and claim (2) separately, since he cannot claim (2). Again, a knave could say: "I am a knave and two plus two is five," but he cannot separately claim: (1) "I am a knave"; (2) "Two plus two is five."

Answer 3: A says that, of A and B, at least one is a knave. If A were a knave, then it would be true that at least one of A and B is a knave and we would have a knave making a true statement, which is not possible. Therefore A must be a knight. Since he is a knight, his statement is true, hence at least one really is a knave. It is then B who must be the knave. So A is a knight and B is a knave.

Answer 4: A is saying that *exactly* one of the persons A and B is a knave. If A is a knight, his statement is true, exactly one is a knave, and so B is a knave. If A is a knave, his statement is false, hence B must again be a knave, because if B were a knight, then it *would* be true that exactly one is a knave! And so regardless of whether A is a knight or a knave, B is a knave. As for A, his type cannot be determined; he could be either a knight or a knave.

Answer 5: If B were a knave, no native would claim to be the same type as B, because that would be tantamount to claiming to be a knave. Therefore B must be a knight, since A *did* claim to be of the same type as B. As for A, he could be either a knight or a knave.

If A had instead said that he and B were *different* types, this would be equivalent to the statement "One of us is a knight and one of us is a knave," which in turn is the same as the statement "Exactly one of us is a knave." This is really the same as Question 4, and so the answer is that B is a knave and A is indeterminate.

Looked at another way, if B were a knight, then no inhabitant would claim to be a *different* type than B!

Now that the review is over, the fun will start!

THE SEARCH FOR ARTHUR YORK

I · THE FIRST TRIAL

Inspector Craig of Scotland Yard—of whom you will read much in this book—was called to the Island of Knights and Knaves to help find a criminal named Arthur York. What made the process difficult was that it was not known whether Arthur York was a knight or a knave.

One suspect was arrested and brought to trial. Inspector Craig was the presiding judge. Here is a transcript of the trial:

CRAIG: What do you know about Arthur York?

DEFENDANT: Arthur York once claimed that I was a knave.

CRAIG: Are you by any chance Arthur York?

DEFENDANT: Yes.

Is the defendant Arthur York?

2 · THE SECOND TRIAL

Another suspect was arrested and brought to trial. Here is a transcript of the trial:

CRAIG: The last suspect was a queer bird; he actually claimed to be Arthur York! Did *you* ever claim to be Arthur York?

DEFENDANT: No.

CRAIG: Did you ever claim that you are *not* Arthur York?

DEFENDANT: Yes.

Craig's first guess was that the defendant was not Arthur York, but are there really sufficient grounds for acquitting him?

3 · THE THIRD TRIAL

"Don't despair," said Craig to the chief of the island police, "we may find our man yet!"

Well, a third suspect was arrested and brought to trial. He brought with him his defense attorney, and the two made the following statements in court.

DEFENSE ATTORNEY: My client is indeed a knave, but he is not Arthur York.

DEFENDANT: My attorney always tells the truth!

Is there enough evidence either to acquit or convict the defendant?

SOME UNUSUAL KNIGHTS AND KNAVES

The puzzles of this section are quite unlike any of my past puzzles about knights and knaves.

4 · MY FIRST ADVENTURE

Inspector Craig left the island shortly after winding up the case of Arthur York. Two days later, I came to this island looking for adventure.

On the first day I arrived, I met an inhabitant and asked him: "Are you a knight or a knave?" He angrily replied: "I refuse to tell you!" and walked away. That's the last I ever saw or heard of him.

Was he a knight or a knave?

5 · MY SECOND DAY

On the next day I came across a native who made a certain statement. I thought for a moment and said: "You know, if you hadn't made that statement, I could have believed it! Before you said it, I had no idea whether it was true or not, nor did I have any prior knowledge that you are a knave. But now that you have said it, I know that it must be false and that you are a knave."

Can you supply a statement that could fulfill those two conditions? *Note:* The statement "Two plus two is five" won't work; I would have already known that statement to be false before he made it.

6 · THE NEXT DAY

On the next day I came across a native who said: "My father once said that he and I are different types, one a knight and one a knave."

Is it possible that his father really said that?

7 · THE NEXT DAY

On the next day I was in a rather frivolous mood. I passed a native and asked him: "Do you ever answer no to questions?"

He answered me—that is, he said either yes or no—and I then knew for sure whether he was a knight or a knave. Which was he?

The above puzzle occurred to me as a result of a clever story told to me by the mathematician Stanislaw Ulam. Ulam referred to it as a paradox. The story is a true one and refers to a certain past president of the United States.

On television Professor Ulam saw this president address the cabinet on his first day in office. He said to them, in a supercilious tone of voice: "You men are not all yes-men, are you?" They all solemnly replied: "Noooo!"

And so it seems that a person doesn't necessarily have to answer yes to be a yes-man! I am also reminded of a cartoon sent to me by one of my readers. It is a drawing of a tough-looking employer saying to his meek-looking employee: "I hate yes-men, Jenkins; don't you?"

8 · THE SOCIOLOGIST

On the next day, I met a sociologist who was visiting the island. He gave me the following report:

"I have interviewed all the natives of this island and I have observed a curious thing: For every native X, there is at least one native Y such that Y claims that X and Y are both knaves."

Does this report hold water?

9 · MY LAST ADVENTURE

My last adventure on this island during that particular visit was a curious one. I met a native who said, "This is not the first time I have said what I am now saying." Was the native a knight or a knave?

SOLUTIONS

1 • If the defendant is Arthur York, we get the following contradiction. Suppose he is Arthur York. Then he is a knight, since he claimed to be Arthur York. That would mean that his first answer to Craig was also true, which means that he, Arthur York, once claimed that he was a knave. But that is impossible! Therefore the defendant is not Arthur York, although he is, of course, a knave.

2 • The defendant is either a knight or a knave. Suppose he is a knight. Then his answers were both truthful; in particular, his second answer was truthful, so he did once claim that he is not Arthur York. His claim was true, since he is a knight; thus he is not Arthur York. This proves that if he is a knight, then he is not Arthur York.

Suppose he is a knave. Then his answers were both lies; in particular, his first answer was a lie, which means that he *did* once claim to be Arthur York. But since he is a knave, he lied when he claimed to be Arthur York, hence he is not Arthur York. And so we have proved that if he is a knave, then he is not Arthur York.

We now see that regardless of whether he is a knight or a knave, he cannot be Arthur York. And so he was acquitted. Incidentally, it cannot be determined whether he is a knight or a knave.

3 • It was very stupid for the defendant to say what he did! Of all the false statements he could have made, he chose just about the most incriminating one possible. Here is why the defendant must be Arthur York:

Suppose the defense attorney is a knight. Then his statement is true, which implies that the defendant is a knave.

Hence the defendant's statement is false, which means that the defense attorney is a knave. So if the defense attorney is a knight, he is also a knave, which is impossible. Therefore the defense attorney can't be a knight; he must be a knave. It then follows that the defendant is also a knave, since he falsely claimed that his attorney always tells the truth. And so we now know that both the attorney and the defendant are knaves. Now, if the defendant were not Arthur York, then it would be true that the defendant is a knave but not Arthur York, hence the attorney would have made a true statement. But the attorney is a knave and *can't* make a true statement! Therefore the defendant must be Arthur York.

4 • He said that he refused to tell me, and by Gad he did refuse! So he told the truth; hence he was a knight.

5 • There are many possible statements that would work. What actually happened was this:

Before he spoke, I had no idea whether he was a knight or a knave, nor did I know whether he was wealthy or not. But then he said: "I am a wealthy knave." A knight could never say he was a wealthy knave, hence I realized that he must be a knave, but not a wealthy one.

A clever alternative solution suggested to me by a bright high school student is: "I am mute."

6 • If he hadn't *said* that his father once said that they were different types, then it would have been possible for his father to have said it. But suppose the father had really said that. Then the father must be a knight and the son must be a knave—see Question 5 and its answer, at the beginning of this chapter—and therefore the son would never have truthfully said that his father said that.

Incidentally, the statement "My father once said that he and I are of different types" provides yet another solution to Problem 5.

7 • If he had answered yes, I would have concluded that he was *probably* a knight. But he answered no, and so I knew for sure that he was a knave, because he just answered no, thus falsely denying that he ever answered no!

8 • If the report is true, we get the following contradiction. For every X there is some Y who claims that X and Y are both knaves. Now, the only way that Y can claim that X and Y are both knaves is that Y is a knave and X is a knight. Therefore every inhabitant X of the island must be a knight. Yet for every inhabitant X there is at least one inhabitant Y who is a knave, since he claims that X and Y are both knaves. So there is at least one knave Y on the island. This contradicts the already proved fact that all the inhabitants are knights.

9 • The native's statement is to the effect that he has made that very statement some time before. Suppose he is a knight. Then he really has made that statement before—say, yesterday. When he made the statement yesterday, he was also a knight, and so it was true, which means that he had made that same statement some time before that—say, the day before yesterday. We thus get an infinite regress; the only way the native can be a knight is if he has lived infinitely back into the past. Therefore the native is really a knave.

Another way of looking at the problem, which some readers will find simpler, is this: Since the native has made the statement once, there must have been a first time he made it. Well, when he made it the first time, it was clearly false, hence he is a knave (and he can never make the same statement again, for it would be true).

CHAPTER 6

DAY-KNIGHTS AND NIGHT-KNIGHTS

We will return to the Island of Knights and Knaves in a later chapter. Meanwhile I would like to tell you of an equally strange place called Subterranea. It is a city completely underground; the inhabitants have never seen the light of day. Clocks, watches, and all other timepieces are strictly forbidden. Yet the inhabitants have an uncanny sense of time; they always know when it is day and when it is night. Each inhabitant is of one of two types—*day-knights* and *night-knights*. The day-knights tell the truth during the day and lie during the night; the night-knights tell the truth during the night and lie during the day.

Visitors to the city are allowed, but of course they may not bring any timepieces with them. Any visitor to the city is bound to become disoriented; after a few days he loses all sense of when it is day and when it is night.

1 · HOW MANY QUESTIONS?

Suppose that you visit this city and after a few days, you lose all sense of time. At one point you would like to know whether it is day or night. You meet one of the inhabitants and are allowed to ask him as many questions as you like, but you may ask only questions whose answer is yes or no. What is the minimum number of questions you need ask to find out whether it is day or night?

· 2 ·

Suppose that instead of wanting to know whether it is day or night, you want to know whether the inhabitant to whom you are speaking is a day-knight or a night-knight. What is the smallest number of questions you need ask?

· 3 ·

When I visited this city, I also lost my time orientation after a few days. Once I came across an inhabitant who made a statement. Before he spoke, I did not know whether he was a day-knight or a night-knight, or whether it was day or night. After he made the statement, I knew that he was a day-knight and that it was then night. Can you supply such a statement?

· 4 ·

On another occasion I came across an inhabitant who made a statement from which I could deduce that he was a day-knight and that it was then day. What statement would work?

· 5 ·

On another occasion I met an inhabitant who said: "During the daylight hours I claim it is night." Was it then day or night?

· 6 ·

On another occasion a native said to me: "During the day I claim that I am a night-knight. I am really a day-knight."

I was happy that he made this statement because I could then deduce both his type and whether it was night or day. What is the solution? *Note:* In this and all other problems in this chapter, I am making the underlying assumption that it never changed from day to night or from night to day during the course of the conversation.

· 7 ·

On one occasion an inhabitant said: "I am a night-knight and it is now day." Was he a day-knight or a night-knight? Was it then day or night?

· 8 ·

On another occasion I asked an inhabitant two questions: "Are you a day-knight?" and "Is it now day?" He replied: "Yes is the correct answer to at least one of your questions."

Was he a day-knight or a night-knight? Was it then day or night?

· 9 ·

I once asked an inhabitant: "Is it true that twelve hours ago you claimed that you were a night-knight?" He replied: "No." I then asked him: "Twelve hours ago, did you claim that you were a day-knight?" He replied: "Yes."

Was he a day-knight or a night-knight? *Note:* I am, of course, assuming that twelve hours makes the difference between day and night.

THE DIFFICULTIES DOUBLE

The next few problems that I encountered presented more of a challenge.

10 · TWO BROTHERS

I once came across two brothers A and B and did not know the type of either, nor did I even know whether they were the same type. I also did not know whether it was day or night at the time. Here is what they said:

A: At least one of us is a day-knight.

B: A is a night-knight.

I then knew the type of each and whether it was night or day. What is the solution?

· II ·

On another occasion I came across two inhabitants A and B who made the following statements:

A: Both of us are day-knights.

B: That is not true!

Which one should be believed?

· 12 ·

At last I made friends with one of the inhabitants—Jim Hawkins—whom I knew to be a day-knight. At one point he told me that some time earlier he had overheard a conversation between two inhabitants A and B in which A said that B was a day-knight and B said that A was a night-knight.

Was it during the day or during the night that Jim told me this?

A METAPUZZLE

· 13 ·

Inspector Craig of Scotland Yard also visited this city. Like every other visitor, he lost his sense of time after a few days. At one point he was desirous of knowing whether it was day or night. He came across a married couple, but he didn't know whether the husband and wife were the same type or different types. One of the two said: "My spouse and I are different types; one of us is a day-knight and one of us is a night-knight." Inspector Craig thought about this and said: "What I really want to know is whether it is now day or night. Which is it?" One of the two said: "It is now day." Inspector Craig then knew whether it was day or night. Was it day or night?

SOLUTIONS

1 • There is a general principle that will prove helpful in many of these problems—namely that during the day, all inhabitants claim to be day-knights, and during the night, all inhabitants claim to be night-knights. This is true because during the day a day-knight is truthful and will truthfully admit to being a day-knight, whereas a night-knight will lie and also claim to be a day-knight. During the night, a night-knight will truthfully claim he is a night-knight and a day-knight will falsely claim he is a night-knight.

Therefore, to find out whether it is day or night, you need ask but one question: Are you a day-knight? If you get yes for an answer, then it is day; if you get no for an answer then it is night.

2 • Another useful principle is that day-knights always claim it is day and night-knights always claim it is night. This is because during the day, a day-knight will truthfully claim it is day and during the night he will falsely claim it is day. On the other hand, during the day, night-knights falsely claim it is night, and during the night, they truthfully claim it is night.

Therefore, to find out if someone is a day-knight or a night-knight, just ask him whether it is now day. If he answers yes, then he is a day-knight; if he answers no, then he is a night-knight.

3 and 4 • The solutions will emerge as a result of some subsequent problems.

5 • He claims that he lies during the day, or what is the same thing, that he is a night-knight. Therefore it must have been night when he said that.

6 • The native's first statement was simply a lie—no one ever claims to be a night-knight during the day, as was explained in the solution to Problem 1. Therefore he was lying at the time, so his second statement was also a lie. Therefore he is really a night-knight. Since he is a night-knight and was lying at the time, it must have been day.

7 • Suppose his statement were true. Then he would really be a night-knight and it would really be day, but night-knights don't tell the truth during the day. Therefore his statement must have been false. So he was lying, but since he is not a night-knight making the statement during the day, it must be that he is a day-knight and that it was then night.

This, by the way, provides a solution to Problem 3.

8 • He is in effect asserting that either he is a day-knight or that it is now day, and maybe both. Suppose his statement were false. Then he is neither a day-knight, nor is it day; this means that he is a night-knight and it is night. But night-knights don't make false statements at night, so it is contradictory to assume that his statement was false. And so his statement was true. Therefore he is either a day-knight or it is day. If the first alternative holds—that is, if he is a day-knight—then it must be day, because day-knights tell the truth only during the day. If the second alternative holds—that is, it is now day—then he must be a day-knight, because only day-knights tell the truth during the day. And so each of the alternatives implies the other, which means that he is a day-knight *and* it is day.

This, incidentally, provides a solution to Problem 4; he could have said: "Either I am a day-knight or it is now day." An equally valid solution is: "If I am a night-knight then it is now day."

9 • His answers were either both truthful or both lies.

Case 1: Both answers were truthful. Then his second answer was truthful, hence twelve hours ago he really did claim to be a day-knight. He lied twelve hours ago, since he is now truthful, so he is really a night-knight.

Case 2: Both answers were lies. Then his first answer was a lie, hence twelve hours ago he *did* claim to be a night-knight. He was truthful then, so he is a night-knight.

In either case he is a night-knight. It cannot be determined whether his answers were truthful or lies.

10 • If it is night, we get the following contradiction: Suppose B is telling the truth. Then A is really a night-knight, hence A is telling the truth, since it is night, which means that at least one of them is a day-knight, hence B must be the day-knight and we have the impossibility of a day-knight telling the truth during the night. Suppose, on the other hand, that B is lying. Then A is really a day-knight, hence his statement is true— that is, at least one of them is a day-knight—which means that A, a day-knight, is telling the truth during the night. This is equally impossible. Therefore we know for sure that it is now day.

Could B be a day-knight? If he were, then he is telling the truth, since it is day, hence A would be a night-knight, but then his statement would be correct, since at least one of them—namely B—is a day-knight, which would mean that A, a night-knight, is making a true statement during the day. Therefore B cannot be a day-knight; he must be a night-knight. Then, since it is now day, his statement is false, hence A is really a day-knight.

And so the solution is that A is a day-knight, B is a night-knight, and it is now day.

11 • This is quite simple: If A were telling the truth, both would be day-knights; B would be the same type as A and wouldn't have contradicted A. And so A is lying and B is telling the truth.

12 • Jim's story cannot be true for the following reasons. Suppose A and B made their statements during the day. If A is a day-knight, then he is telling the truth, hence B is a day-knight and his statement is true, but B said that A is a night-knight, so B's statement can't be true. This is a contradiction. On the other hand, if A is a night-knight, then A is lying, which means that B is a night-knight. But since A is a night-knight, then B's statement was truthful, which means that B, a night-knight, told the truth during the day. This proves that the remarks of A and B couldn't have occurred during the day. A symmetrical argument, which the reader can supply, shows that A and B couldn't have made these remarks during the night either. Therefore Jim's story was simply false. Since Jim is a day-knight, he must have told this story during the night.

13 • Let A be the one—husband or wife, as the case may be—who claimed that his or her spouse B was not the same type as A. If A's statement was true, then A was then in the truth-telling state and B was really a different type than A, hence B was then in the lying state. If A's statement was false, then A was in the lying state, B was actually the same type as A, and again B had to be in the lying state. So Craig could deduce from A's statement that B was in the lying state, but Craig could not know whether or not A was in the truth-telling state. Therefore, if A had been the one who claimed it was day, Craig couldn't have known whether it was day or night. And so it must have been B who claimed it was day, and Craig then knew it was night.

CHAPTER 7

GODS, DEMONS, AND MORTALS

Shortly after Inspector Craig returned to London from his strange experience in Subterranea, he had a curious dream. He had been browsing that day in a library specializing in rare books on mythology, another of his many interests. His head was filled with gods and demons, and so his dream was perhaps not so surprising.

Time sometimes passes in unusual ways in the course of a dream. Craig dreamed that he spent nine days in a region in which dwelled gods, demons, and mortals. The gods, of course, always told the truth, and the demons always lied. As to the mortals, half were knights and half were knaves. As usual, the knights told the truth and the knaves lied.

1 · THE FIRST DAY

Craig dreamed that on the first day he met a dweller of the region who looked as if he might be a god, though Craig could not be sure. The dweller evidently guessed Craig's thoughts, smiled, and made a statement to reassure him. From this statement, Craig *knew* that he was in the presence of a god. Can you supply such a statement?

2 · THE SECOND DAY

In this episode of the dream, Craig met a terrifying-looking being who had every appearance of being a demon.

52

"What sort of being are you?" asked Craig, in some alarm.

The being answered, and Craig then realized that he was confronting not a demon, but a knave. What could the being have answered?

3 · THE THIRD DAY

In this episode, Craig met a totally nondescript-looking being who from appearances could have been anything at all. The being then made a statement from which Craig could deduce that he was either a god or a demon, but Craig could not tell which.

Can you supply such a statement?

4 · THE FOURTH DAY

Craig next met a being who made the following two statements:

1. A god once claimed that I am a demon.
2. No knight has ever claimed that I am a knave.

What sort of being was he?

5 · THE FIFTH DAY

A being made the following two statements to Craig:

1. I never claim to be a knave.
2. I sometimes claim that I am a demon.

What sort of being are we now dealing with?

6 · THE SIXTH DAY

In this episode, Craig came across *two* beings, each of whom made a statement. Craig could then infer that at least one of them must be a god, but he could not tell which one. From neither statement alone could Craig have deduced this. What statements could the beings have made?

7 · THE SEVENTH DAY

On the next day, Craig again met two beings each of whom made a statement. Craig could then infer that one of them was a knave and the other a demon, though he could not tell which was which. Again, from neither statement alone could Craig have inferred this. Can you supply two such statements?

8 · INTRODUCING THOR

On the eighth day, Craig met a being who had every appearance of being the god Thor. The being made a statement, and Craig then knew he must *be* Thor. What statement could Thor have made?

9 · A PERPLEXITY RESOLVED

Craig and Thor became fast friends. In fact, on the evening of the ninth day, Thor gave a magnificent banquet in Craig's honor. "I propose a toast to our illustrious guest!" said Thor, as he raised his glass of nectar.

After a round of cheers, Craig was asked to speak.

"I am very perplexed!" said Craig as he rose. "I wonder if this may not all be a dream!"

"Why do you think you may be dreaming?" asked Thor.

"Because," said Craig, "two incidents have occurred today that seem totally inexplicable. This morning I met someone who made a statement which no knight, knave, god, or demon could possibly make. Then this afternoon I met someone else who also made a statement which no dweller of this region could possibly make. That is why I suspect that I may be dreaming."

"Oh!" said Thor. "Be reassured; you are not dreaming. The two incidents have a perfectly rational explanation. You see, we have had two visitors here from another realm. Both of them are mortal. One is Cyrus, who always tells the truth, although he is not called a knight since he is not from this region. The other is Alexander, who sometimes tells the truth and sometimes lies. It must have been those two whom you met today. What statements did they make?"

Craig then told the company what each had said.

"That explains it perfectly!" said Thor. "Moreover, it follows from their having said what they did that Cyrus was the one you met in the morning. And interestingly enough, if you hadn't met Alexander in the afternoon, you could never have known whether the one you met in the morning was Cyrus or Alexander."

Craig thought the matter over and realized that Thor was right.

What statements could these two outsiders have made which fulfill all of the above conditions?

· EPILOGUE—A PHILOSOPHICAL PUZZLE ·

The next morning when Craig was wide awake and recalling his dream, he wondered whether he had been logically inconsistent in his sleep. "The trouble is this," thought Craig: "In my dream I believed that Thor was a god and that gods always tell the truth. Yet Thor told me that I wasn't dreaming. Now how could Thor, who tells the truth, say that I wasn't dreaming when in fact I was? Wasn't this an inconsistency on my part?"

Would you say that Craig's dream was logically inconsistent?

SOLUTIONS

1 · One statement that works is: "I am not a knight." If the speaker were a knave or a demon, then it would be true that he was not a knight, but knaves and demons don't make true statements. Therefore the speaker was neither a knave nor a demon, hence he was a knight or a god and his statement was true. Since it was true, then he really is not a knight; hence he must be a god.

2 · A statement that works is: "I am a demon." Obviously no demon can claim he is a demon, so the speaker is not a demon. Therefore his statement was false and since he is not a demon, he must be a knave.

Incidentally, this and the last puzzle are essentially the same as Problems 4 and 5 of Chapter 1, the puzzles about the prizes.

3 • This is a bit more tricky: A statement that works is: "I am either a god or a knave." That could be said by a god, since a god is either a god or a knave; it could also be falsely said by a demon. It couldn't be said by a knight, because a knight would never lie and claim that he is either a god or a knave, and it couldn't be said by a knave, because a knave would never admit to the true fact that he is either a god or a knave. And so the speaker must be either a god or a demon, but there is no way to tell which.

4 • The speaker's first statement was obviously false, for if it were true, a god would have once claimed that the speaker was a demon, which would mean that the speaker really was a demon, but no one who tells the truth can be a demon. Since the first statement was false, so was the second statement, since it was made by the same speaker. Therefore a knight *did* once claim that the speaker was a knave, hence the speaker really is a knave.

5 • The speaker's second statement was obviously a lie, be cause no truthteller would ever say that he sometimes claims to be a demon. Therefore the first statement was also a lie, hence the speaker *does* sometimes claim to be a knave, hence he must be a demon.

6 • Many solutions are possible; here is one. Let us call the two beings A and B. Now, suppose A and B make the following two statements:

A: B is a knight.

B: A is not a knight.

A is either telling the truth or lying.

Case 1—A is telling the truth: Then B really is a knight, hence his statement is true, hence A is not a knight, therefore A must be a god, since he is telling the truth.

Case 2—A is lying: Then B is not a knight, since A says he is. Also, since A is lying, then A is certainly not a knight, hence B's statement is true. Therefore B is telling the truth, but is not a knight, hence B is a god.

So if Case 1 is true, A is a god; if Case 2 is true, then B is a god. There is no way to tell whether A is telling the truth or lying.

7 • Again let us call the two beings A and B. The following statements would work:

A: Both of us are knaves.

B: Both of us are demons.

It is obvious that both are lying. Since A is lying, they are not both knaves. Since B is lying, they are not both demons. Therefore one is a knave and one is a demon, but there is no way to tell which one is which.

8 • A statement that works is: "I am either a knave or a demon or the god Thor."

If the speaker were either a knave or a demon, then it would be true that he is either a knave or a demon or the god Thor. This would mean that a knave or a demon made a true statement, which is not possible. Therefore the speaker is neither a knave nor a demon, hence his statement is true. Hence he must be the god Thor.

9 • Here is one possible solution.

MORNING SPEAKER: "I am neither a knight nor a god."

AFTERNOON SPEAKER: "I am either a knave or a demon."

No inhabitant of the region could make either of those statements. No knight or god could claim that he is neither a knight nor a god; no knave or demon could make the true statement that he is neither a knight nor a god. As for the second statement, obviously no knight or god would claim to be either a knave or a demon and no knave or demon would admit to being a knave or a demon. Therefore both were outsiders; namely, Cyrus and Alexander. The statement of the morning speaker was true and the statement of the afternoon speaker was false. Since Cyrus never makes false statements, he couldn't have been the afternoon speaker. Thus he was the morning speaker.

· DISCUSSION OF THE EPILOGUE ·

As I see it, Craig's dream was *not* necessarily inconsistent. If Craig had actually *believed* in the dream that he was dreaming, then the set of his beliefs during his dream would have been inconsistent, since the following propositions are indeed logically contradictory: (1) Thor is a god; (2) Gods make only true statements; (3) Thor stated that Craig was not dreaming; (4) Craig was dreaming.

The contradiction is obvious. However, there is no evidence that Craig at any time of his dream believed that he *was* dreaming, although at one point he wondered whether he *might* be dreaming. Craig presumably believed that he was awake, and this belief, though false, was perfectly consistent with the other beliefs of his dream.

Curiously enough, if Craig had formulated the belief that he *was* dreaming, then this belief, though correct, would have created a logical inconsistency!

CHAPTER 8

In Search of the Fountain of Youth

Introduction

Arthur Reynolds, Esq., was in search of a recipe for immortality. He read a great deal of occult and alchemical literature but could find nothing of practical value. Then he heard of a great Sage of the East who was reputed to be a specialist in this area. At great expense, he made a lengthy journey and finally found the Sage.

"Is it really possible to live forever?" he asked the Sage.

"Oh, quite easily," replied the Sage, "provided you do just two things."

"What are the two things?" asked Reynolds eagerly.

"First of all you must never make false statements; in fact, you must resolve to make only true statements from now on. That's a small price to pay for immortality, isn't it?"

"Oh, definitely!" replied Reynolds. "But what about the second thing?"

"The second thing is that you say now: 'I will repeat this statement tomorrow.' If you do just these two things," concluded the Sage, "then I guarantee that you will live forever!"

A Question for the Reader: Is it really true that if you do these two things you will live forever? The answer is given in the text that follows, but the reader might like to think about this before reading further.

Reynolds thought about this for a bit. "Oh, of course!" he said suddenly. *"If* I do those two things, then I will certainly live forever, because if I truthfully say now, 'I will repeat this sentence tomorrow,' then tomorrow I will again say, 'I will repeat this sentence tomorrow,' and if I am truthful tomorrow, then I will say the same thing again the next day, and so on throughout all eternity."

"Exactly!" said the Sage with a triumphant smile.

"But the solution is not a practical one!" protested Reynolds. "How can I truthfully say that I will do something tomorrow if I don't know for sure whether I'll even be alive tomorrow?"

"Oh, you want a *practical* solution," said the Sage. "I didn't realize that. No, I'm not very good at practical solutions; I deal mainly with theory. But a *practical* solution? The only thing I can think of is the Fountain of Youth. Have you considered searching for it?"

"The Fountain of Youth?" cried Reynolds incredulously. "Why, I've read about it in history books, and I know that many have sought it, but does it exist in reality or only in the imagination?"

"That I do not know," replied the Sage, "but if it does exist, I know as likely a place as any where it can be found."

"What place is that?" asked Reynolds.

"The Island of Knights and Knaves," replied the Sage. "I cannot guarantee that the fountain is there, but if it is *anywhere,* that island is as likely a place as any."

Reynolds thanked the Sage and forthwith departed for the Island of Knights and Knaves.

In Search of the Fountain of Youth

And so we are back to the Island of Knights and Knaves. Reynolds arrived without mishap, and his adventures began.

1 · A Preliminary Incident

On the first day, Reynolds met a native who made a statement. Reynolds then realized that if the native was a knight, the Fountain of Youth must be on the island, but if the native was a knave, there was no way of telling whether or not the Fountain of Youth was on the island.

What statement could the native have made?

2 · A Grand Metapuzzle

Reynolds's next adventure was far more interesting—it involves as profound a logic puzzle as any I have ever come across.

On the following day, Reynolds came across two natives A and B and said to them, "Please tell me whatever you know about the Fountain of Youth. Is it on this island?" The two natives then made the following statements:

A: If B is a knave, then the Fountain of Youth is on this island.

B: I never claimed that the Fountain of Youth is not on this island!

Reynolds thought about this for a bit and said, "Please, now, I want a *definite* answer! Is the Fountain of Youth on this island?" One of the two answered—he either said yes or he said no—and Reynolds then knew whether or not the Fountain of Youth was on the island.

Some months later, Reynolds told the above facts to Inspector Craig. (He was a good friend of Craig's and knew of his fondness for logic puzzles.) Craig said, "It is obviously impossible for me to deduce from the facts you have given me whether or not the Fountain of Youth is on this island. You haven't told me whether A or B was the one who answered your second question, or what answer he gave. Whichever one it was who answered, suppose the other one had answered instead. Do you know if you could then have decided whether or not the fountain is on the island?"

Reynolds thought about this and finally told Craig whether or not he knew whether he could have decided had the other one answered the second question instead.

"Thank you," said Craig. "I now know whether or not the Fountain of Youth is on the island."

Is the Fountain of Youth on the island?

SOLUTIONS

1 • One possibility is that the native said: "I am a knight and the Fountain of Youth is on this island." If he was a knight, then the fountain is obviously on the island. If he was not a knight, then what he said is false, regardless of whether or not the fountain is on the island, and there would be no way of telling whether or not the fountain was there.

2 • B said that he had never claimed that the Fountain of Youth was not on the island. If B is a knave, then B *has* previously claimed that the Fountain of Youth is not on the island, and since he is a knave, the fountain *is* on the island! So we now know that if B is a knave, the Fountain of Youth is on the island. Well, this is just what A said; therefore A must be a knight. And so we now know the following two facts:

Fact 1: A is a knight.

Fact 2: If B is a knave, the Fountain of Youth is on the island.

Of course Reynolds, who can reason as well as you and I, realized these two facts also.

Now, we are not told who answered Reynolds's second question, or whether the answer was yes or no, and so there are four possible cases that we must analyze.

Case A_1: A claimed that the fountain was on the island (by answering yes).

In this case, Reynolds, knowing that A is a knight, would know that the fountain was on the island.

Case A_2: A claimed that the fountain was not on the island (by answering no).

In this case, Reynolds would have known that the fountain was not on the island.

Case B_1: B claimed that the fountain was on the island (by answering yes).

In this case, Reynolds would know that the fountain was on the island by reasoning as follows: "Suppose the fountain is not on the island. Then B is a knave for having just affirmed that it is. But by Fact 2, if B is a knave, the fountain *is* on the island. This is a contradiction, hence the fountain must be on the island after all (and also B must be a knight)."

Case B_2: B claimed that the fountain was not on the island (by answering no).

In this case, Reynolds could not possibly know whether or not the fountain was on the island; B could be a knave who falsely claimed that the fountain was not on the island and who also falsely claimed that he never claimed that the fountain was not on the island, or he could be a knight who truthfully claimed that the fountain was on the island and who also claimed that the fountain was not on the island. And so in this case, Reynolds couldn't decide.

However, we are given that Reynolds *did* decide, and therefore Case B_2 is ruled out. So we now know that one of three cases—A_1, A_2, B_1—is the one that holds, and we can henceforth forget about the fourth case B_2.

At this point, we must take into account the second part of the story—the conversation Reynolds had with Inspector Craig. It is important to realize that Craig did not ask Reynolds whether if the other one had answered the question, Reynolds could have decided; Craig asked Reynolds whether he *knew* whether or not he could have decided. Let us see how Reynolds would reason in response to Craig's question. Of course Reynolds *knows* which of the three cases—A_1, A_2, B_1—is the actual one, although *we* don't (at least not yet), and so we must see how Reynolds would reason in each of the three cases.

Case A_1: Reynolds would reason thus: "A is a knight; A said that the fountain is on the island; the fountain is on the island. Now suppose B had answered my second question. I don't know whether he would have answered yes or no. Suppose he had answered yes. Then I would have known that the fountain was on the island, because I would have reasoned that if the fountain were not on the island then B is a knave for claiming it was. But I also know (Fact 2) that if B is a knave, the fountain *is* on the island and so I would have gotten a contradiction from the assumption that the fountain is not on the island. Therefore, if B had answered yes, I would have known that the fountain was on the island. But suppose he had answered no? Then I would have had no way of knowing whether or not the fountain was on the island; I would have reasoned that he could be a knight and the fountain was not on the island, or he could be a knave and the fountain was on the island. And so if he had answered no, then I couldn't have decided whether or not the fountain was on the island.

"In summary, had B answered yes, I could have decided; had he answered no, I couldn't have decided. Since I have no way of knowing what answer B would have given, I have no way of knowing whether I could have decided or not."

Case A_2: In this case, here is how Reynolds would reason: "The fountain is not on this island; A told me this and A is a knight. Now, suppose B had answered instead. Well, B is a knight, because I have already proved that if B were a knave, then the fountain *would* be on this island, which it isn't. Since B is a knight, then had he been the one to answer, he would also have answered no. But then I couldn't have *known* that he was a knight, and so I would have had no way of knowing whether the fountain is on the island or not. In brief, had B been the one to answer, then I definitely could *not* have decided whether or not the fountain was on the island."

Case B_1: This is the simplest case of all! In this case, B is the one who really answered, hence Reynolds would already know that if A had been the one to answer, he could have decided about the fountain, because he already knew that A was a knight.

We now see that if Case A_1 were the actual one, then Reynolds could not know whether he could have decided, hence his answer to Craig would be: "No, I don't know whether I could have decided had the other one been the one who answered my second question."

If Case A_2 is the actual one, then Reynolds would have told Craig: "Yes; I do know whether or not I could have decided." Reynolds, in fact, even knows that he *couldn't* have decided.

If Case B_1 is the actual one, then Reynolds would again have told Craig: "Yes, I know whether or not I could have decided." He in fact even knows that he *could* have decided.

Therefore, if Reynolds answered yes to Craig's question, then either Case A_2 or Case B_1 holds, but there is no way we or Craig could tell which, hence Craig couldn't have known whether or not the Fountain of Youth was on the island. But we are given that Craig *did* know, hence Reynolds must have answered no and Craig then knew that Case A_1 was the only possibility and that the fountain was on the island.

So! The Fountain of Youth is really somewhere on the Island of Knights and Knaves! Finding it, however, is a very different story. In general, it is not easy actually to find things on this crazy island! As a matter of fact, Reynolds didn't *find* the Fountain of Youth on this particular visit; he merely found out for sure that the fountain was somewhere on the island, and he is planning to go back in search of it. But that is a topic for another book.

PART THREE

To Mock a Mockingbird

CHAPTER 9

TO MOCK A MOCKINGBIRD

A certain enchanted forest is inhabited by talking birds. Given any birds A and B, if you call out the name of B to A, then A will respond by calling out the name of some bird to you; this bird we designate by AB. Thus AB is the bird named by A upon hearing the name of B. Instead of constantly using the cumbersome phrase "A's response to hearing the name of B," we shall more simply say: "A's response to B." Thus AB is A's response to B. In general, A's response to B is not necessarily the same as B's response to A—in symbols, AB is not necessarily the same bird as BA. Also, given three birds A, B, and C, the bird A(BC) is not necessarily the same as the bird (AB)C. The bird A(BC) is A's response to the bird BC, whereas the bird (AB)C is the response of the bird AB to the bird C. The use of parentheses is thus necessary to avoid ambiguity; if I just wrote ABC, you could not possibly know whether I meant the bird A(BC) or the bird (AB)C.

Mockingbirds: By a *mockingbird* is meant a bird M such that for any bird x, the following condition holds:

$$Mx = xx$$

M is called a mockingbird for the simple reason that its response to any bird x is the same as x's response to itself—in other words, M *mimics* x as far as its response to x goes. This means that if you call out x to M or if you call out x to itself, you will get the same response in either case.[*]

Composition: The last technical detail before the fun starts is this: Given any birds A, B, and C (not necessarily distinct) the bird C is said to *compose* A with B if for every bird x the following condition holds:

$$Cx = A(Bx)$$

In words, this means that C's response to x is the same as A's response to B's response to x.

TO MOCK A MOCKINGBIRD

I · THE SIGNIFICANCE OF THE MOCKINGBIRD

It *could* happen that if you call out B to A, A might call the same bird B back to you. If this happens, it indicates that A is *fond* of the bird B. In symbols, A is fond of B means that AB = B.

We are now given that the forest satisfies the following two conditions.

C_1 *(the composition condition):* For any two birds A and B (whether the same or different) there is a bird C such that for any bird x, Cx = A(Bx). In other words, for any birds A and B there is a bird C that composes A with B.

C_2 *(the mockingbird condition):* The forest contains a mockingbird M.

One rumor has it that every bird of the forest is fond of at least one bird. Another rumor has it that there is at least one bird that is not fond of any bird. The interesting thing is that it is possible to settle the matter completely by virtue of the given conditions C_1 and C_2.

Which of the two rumors is correct?

Note: This is a basic problem in the field known as *combinatory logic.* The solution, though not lengthy, is extremely ingenious. It is based on a principle that derives ultimately from the work of the logician Kurt

---

[*]For handy reference to the birds, each is alphabetically listed in "Who's Who Among the Birds," p. 231.

Gödel. This principle will permeate parts of many of the chapters that follow.

2 · EGOCENTRIC?

A bird x is called *egocentric* (sometimes *narcissistic*) if it is fond of itself—that is, if x's response to x is x. In symbols, x is egocentric if xx = x.

The problem is to prove that under the given conditions C_1 and C_2 of the last problem, at least one bird is egocentric.

3 · STORY OF THE AGREEABLE BIRD

Two birds A and B are said to *agree* on a bird x if their responses to x are the same—in other words if Ax = Bx. A bird A is called *agreeable* if for every bird B, there is at least one bird x on which A and B agree. In other words, A is *agreeable* if for every bird B there is a bird x such that Ax = Bx.

We now consider the following variant of Problem 1: We are given the composition condition C_1, but we are not given that there is a mockingbird; instead, we are given that there is an agreeable bird A. Is this enough to guarantee that every bird is fond of at least one bird?

A bonus question: Why is Problem 1 nothing more than a special case of Problem 3? *Hint:* Is a mockingbird necessarily agreeable?

4 · A QUESTION ON AGREEABLE BIRDS

Suppose that the composition condition C_1 of Problem 1 holds and that A, B, and C are birds such that C composes A with B. Prove that if C is agreeable then A is also agreeable.

5 · AN EXERCISE IN COMPOSITION

Again suppose that condition C_1 holds. Prove that for any birds A, B, and C there is a bird D such that for every bird x, Dx = A(B(Cx)). This fact is quite useful.

6 · COMPATIBLE BIRDS

Two birds A and B, either the same or different, are called *compatible* if there is a bird x and a bird y, either the same or different, such that Ax = y and By = x. This means that if you call out x to A then you will get y as a response, whereas if you call out y to B, you will get x as a response.

Prove that if conditions C_1 and C_2 of Problem 1 hold, then any two birds A and B are compatible.

7 · HAPPY BIRDS

A bird A is called *happy* if it is compatible with itself. This means that there are birds x and y such that Ax = y and Ay = x.

Prove that any bird that is fond of at least one bird must be a happy bird.

8 · NORMAL BIRDS

We will henceforth call a bird *normal* if it is fond of at least one bird. We have just proved that every normal bird is happy. The converse is not necessarily true; a happy bird is not necessarily normal.

Prove that if the composition condition C_1 holds and if there is at least one happy bird in the forest, then there is at least one normal bird.

HOPELESS EGOCENTRICITY

9 · HOPELESSLY EGOCENTRIC

We recall that a bird B is called *egocentric* if BB = B. We call a bird B *hopelessly egocentric* if for *every* bird x, Bx = B. This means that whatever bird x you call out to B is irrelevant; it only calls B back to you! Imagine that the bird's name is Bertrand. When you call out "Arthur," you get the response "Bertrand"; when you call out "Raymond," you get the response "Bertrand"; when you call out "Ann," you get the response "Bertrand." All this bird can ever think of is itself!

More generally, we say that a bird A is *fixated* on a bird B if for every bird x, Ax = B. That is, all A can think of is B! Then a bird is hopelessly egocentric just in the case that it is fixated on itself.

A bird K is called a *kestrel* if for any birds x and y, (Kx)y = x. Thus if K is a kestrel, then for every bird x, the bird Kx is fixated on x.

Given conditions C_1 and C_2 of Problem 1, and the existence of a kestrel K, prove that at least one bird is hopelessly egocentric.

10 · FIXATION

If x is fixated on y, does it necessarily follow that x is fond of y?

11 · A FACT ABOUT KESTRELS

Prove that if a kestrel is egocentric, then it must be hopelessly egocentric.

12 · ANOTHER FACT ABOUT KESTRELS

Prove that for any kestrel K and any bird x, if Kx is egocentric then K must be fond of x.

13 · A SIMPLE EXERCISE

Determine whether the following statement is true or false: If a bird A is hopelessly egocentric, then for any birds x and y, Ax = Ay.

14 · ANOTHER EXERCISE

If A is hopelessly egocentric, does it follow that for any birds x and y, (Ax)y = A?

15 · HOPELESS EGOCENTRICITY IS CONTAGIOUS!

Prove that if A is hopelessly egocentric, then for every bird x, the bird Ax is also hopelessly egocentric.

16 · ANOTHER FACT ABOUT KESTRELS

In general, it is not true that if Ax = Ay then x = y. However, it *is* true if A happens to be a kestrel K. Prove that if Kx = Ky then x = y. (We shall henceforth refer to this fact as the *left cancellation law for kestrels*.)

17 · A FACT ABOUT FIXATION

It is possible that a bird can be fond of more than one bird, but it is not possible for a bird to be fixated on more than one bird. Prove that it is impossible for a bird to be fixated on more than one bird.

18 · ANOTHER FACT ABOUT KESTRELS

Prove that for any kestrel K and any bird x, if K is fond of Kx, then K is fond of x.

19 · A RIDDLE

Someone once said: "Any egocentric kestrel must be extremely lonely!" Why is this true?

IDENTITY BIRDS

A bird I is called an *identity* bird if for every bird x the following condition holds:

$$Ix = x$$

The identity bird has sometimes been maligned, owing to the fact that whatever bird x you call to I, all I does is to echo x back to you. Superficially, the bird I appears to have no intelligence or imagination; all it can do is repeat what it hears. For this reason, in the past, thoughtless students of ornithology referred to it as the *idiot* bird. However, a more profound ornithologist once studied the situation in great depth and discovered that the identity bird is in fact highly intelligent! The *real* reason for its apparently unimaginative behavior is that it has an unusually large heart *and hence is fond of every bird!* So when you call x to I, the reason it responds by calling back x is not that it can't think of anything else; it's just that it wants you to know that it is fond of x!

Since an identity bird is fond of every bird, then it is also fond of itself, so every identity bird is egocentric. However, its egocentricity doesn't mean that it is any more fond of itself than of any other bird!

Now for a few simple problems about identity birds.

· 20 ·

Supposing we are told that the forest contains an identity bird I and that I is agreeable, in the sense of Problem 3. Does it follow that every bird must be fond of at least one bird? *Note:* We are no longer given conditions C_1 and C_2.

· 21 ·

Suppose we are told that there is an identity bird I and that every bird is fond of at least one bird. Does it necessarily follow that I is agreeable?

· 22 ·

Suppose we are told that there is an identity bird I, but we are not told whether I is agreeable or not. However, we are told that every pair of birds is compatible, in the sense of Problem 6. Which of the following conclusions can be validly drawn?

1. Every bird is normal—i.e., fond of at least one bird.
2. I is agreeable.

23 · WHY?

The identity bird I, though egocentric, is in general not *hopelessly* egocentric. Indeed, if there were a hopelessly egocentric identity bird, the situation would be quite sad. Why?

LARKS

A bird L is called a *lark* if for any birds x and y the following holds:

$$(Lx)y = x(yy)$$

Larks have some interesting properties, as we will now see.

· 24 ·

Prove that if the forest contains a lark L and an identity bird I, then it must also contain a mockingbird M.

· 25 ·

One reason I like larks is this: If there is a lark in the forest, then it follows without further ado that every bird is fond of at least one bird. And so you see, the lark has a wonderful effect on the forest as a whole; its presence makes every bird normal. And since all normal birds are happy, by Problem 7, then a lark L in the forest causes all the birds to be happy! Why is this true?

26 · ANOTHER RIDDLE

Why is a hopelessly egocentric lark unusually attractive?

· 27 ·

Assuming that no bird can be both a lark and a kestrel—as any ornithologist knows!—prove that it is impossible for a lark to be fond of a kestrel.

· 28 ·

It might happen, however, that a kestrel K is fond of a lark L. Show that if this happens, then every bird is fond of L.

· 29 ·

Now let me tell you the most surprising thing I know about larks: Suppose we are given that the forest contains a lark L and we are not given any other information. From just this one fact alone, it can be proved that at least one bird in the forest must be egocentric!

The proof of this is a bit tricky. Given the lark L, we can actually write down an expression for an egocentric bird—and we can write it using just the letter L, with parentheses, of course. The shortest expression that I have been able to find has a length of 12, not counting parentheses. That is, we can write L twelve times and then by parenthesizing it the right way, have the answer. Care to try it? Can you find a shorter expression than mine that works? Can it be proved that there is no shorter expression in L that works? I don't know! At any rate, see if you can find an egocentric bird, given the bird L.

SOLUTIONS

1 • The first rumor is correct; every bird A is fond of at least one bird. We prove this as follows:

Take any bird A. Then by condition C_1, there is a bird C that composes A with the mockingbird M, because for *any* bird B, there is a bird C that composes A with B, so this is also true if B happens to be the mockingbird M. Thus for any bird x, $Cx = A(Mx)$, or what is the same thing, $A(Mx) = Cx$. Since this equation holds for *every* bird x, then we can substitute C for x, thus getting the equation $A(MC) = CC$.

But $MC = CC$, since M is a mockingbird, and so in the equation $A(MC) = CC$, we can substitute CC for MC, thus getting the equation $A(CC) = CC$. This means that A is fond of the bird CC!

In short, if C is any bird that composes A with M, then A is fond of the bird CC. Also, A is fond of MC, since MC is the same as the bird CC.

2 • We have just seen that conditions C_1 and C_2 imply that every bird is fond of at least one bird. This means, in particular, that the mockingbird M is fond of at least one bird E. Now we show that E must be egocentric.

First, $ME = E$, since M is fond of E. But also $ME = EE$, because M is a mockingbird. So E and EE are both identical with the bird ME, so $EE = E$. This means that E is fond of E—i.e., that E is egocentric.

Remark: Since E is egocentric and E = ME, then ME is egocentric. Doesn't the word "ME" tell its own tale?

3 • We are given that the composition condition C_1 holds and that there is an agreeable bird A.

Take any bird x. By the composition condition, there is some bird H that composes x with A. Since A is agreeable, then A agrees with H on some bird y. We will show that x must be fond of the bird Ay.

Since A agrees with H on y, then Ay = Hy. But since H composes x with A, then Hy = x(Ay). Therefore Ay = Hy = x(Ay), and so Ay = x(Ay), or what is the same thing, x(Ay) = Ay. This means that x is fond of Ay.

A bonus question: The mockingbird is certainly agreeable, because for any bird x, M agrees with x on the very bird x, since Mx = xx. In other words there *is* a bird y—namely x itself—such that My = xy.

Since every mockingbird is agreeable, then the given conditions of Problem 3 imply the given conditions of Problem 1, and therefore the solution of Problem 1 gives an alternative solution to Problem 3, though a more complicated one.

4 • We are given that C composes A with B and that C is agreeable. We are also given the composition condition. We are to show that A is agreeable.

Take any bird D. We must show that A agrees with D on some bird or other. Since the composition law holds, then there is a bird E that composes D with B. Also C agrees with E on some bird x, because C is agreeable—thus Cx = Ex. Also Ex = D(Bx), because E composes D with B, and Cx = A(Bx), because C composes A with B. Therefore, since Ex = D(Bx), we have A(Bx) = D(Bx). And so A agrees with D on the bird Bx. This proves that for *any* bird D, there is a bird on which A and D agree, which means that A is agreeable.

In short, A(Bx) = Cx = Ex = D(Bx).

5 • Suppose the composition law C_1 holds. Take any birds A, B, and C. Then there is a bird E that composes B with C, and so for any bird x, Ex = B(Cx), and hence A(Ex) = A(B(Cx)). Using the composition law again, there is a bird D that composes A with E, and hence Dx = A(Ex). Therefore Dx = A(Ex) = A(B(Cx)), and so Dx = A(B(Cx)).

6 • We are given that conditions C_1 and C_2 of Problem 1 hold. Therefore every bird is fond of at least one bird, ac cording to the solution to Problem 1. Now take any birds A and B. By condition C_1, there is a bird C that composes A with B. The bird C is fond of some bird—call it y. Thus Cy = y. Also Cy = A(By)—because C composes A with B. Therefore A(By) = y. Let x be the bird By. Then Ax = y, and of course By = x. This proves that A and B are compatible.

7 • To say that A is compatible with B doesn't necessarily mean that there are two *distinct* birds x and y such that Ax = y and By = x; x and y may be the same bird. So if there is a bird x such that Ax = x and Bx = x, that surely implies that A and B are compatible. Thus if Ax = x, then A is automatically compatible with A, because Ax = y and Ay = x, when y is the same bird as x.

Therefore if A is fond of x, then Ax = x, and A is compatible with A, which means that A is happy.

8 • Suppose H is a happy bird. Then there are birds x and y such that Hx = y and Hy = x. Since Hx = y, then we can substitute Hy for x (since Hy = x) and obtain H(Hy) = y. Also, by the composition condition C_1, there is a bird B that composes H with H, and so By = H(Hy) = y. So By = y, which means that B is fond of y. Since B is fond of some bird y, then B is normal.

9 • We are given the conditions of Problem 1, hence every bird is fond of at least one bird. In particular, the kestrel K is fond of some bird A. Thus KA = A. Hence for every bird x, (KA)x = Ax. Also (KA)x = A, since K is a kestrel. Therefore Ax = A. Since for every bird x, Ax = A, then A is hopelessly egocentric.

We can also look at the matter this way: If the kestrel K is fond of a bird A, then KA = A. Also KA is fixated on A, and since KA = A, then A is fixated on A, which means that A is hopelessly egocentric. And so we see that any bird of which the kestrel is fond must be hopelessly egocentric.

10 • Of course it does! If x is fixated on y, then for *every* bird z, xz = y, hence in particular, xy = y, which means that x is fond of y.

11 • If K is egocentric, then K is fond of K. But we proved in Problem 9 that any bird of which K is fond must be hopelessly egocentric, and so if K is egocentric, K is hopelessly egocentric.

12 • Suppose that Kx is egocentric. Then (Kx)(Kx) = Kx. But also (Kx)(Kx) = x, because for *any* bird y, (Kx)y = x, so this is also true when y is the bird Kx. Therefore Kx = x, because Kx and x are both equal to the bird (Kx)(Kx), so they are equal to each other. This means that K is fond of x.

13 • Suppose A is hopelessly egocentric. Then Ax = A and Ay = A, so Ax = Ay; they are both equal to A. Thus the statement is true.

14 • Yes, it does follow. Suppose A is hopelessly egocentric. Then Ax = A, hence (Ax)y = Ay and Ay = A, so (Ax)y = A.

15 • Suppose A is hopelessly egocentric. Then for any birds x and y, (Ax)y = A, according to the last problem. Also Ax = A, since A is hopelessly egocentric. Therefore (Ax)y = Ax, since (Ax)y and Ax are both equal to A. Therefore, for any bird y, (Ax)y = Ax, which means that Ax is hopelessly ego centric.

16 • Suppose Kx = Ky and that K is a kestrel. Then for any bird z, (Kx)z = (Ky)z. But (Kx)z = x and (Ky)z = y, so x = (Kx)z = (Ky)z = y. Therefore x = y.

17 • Suppose A is fixated on x and A is fixated on y; we will show that x = y.

Take any bird z. Then Az = x, since A is fixated on x, and Az = y, since A is fixated on y. Therefore x and y are both equal to the bird Az, and so x = y.

18 • Suppose K is fond of Kx. Then K(Kx) = Kx. Now, K(Kx) is fixated on Kx, whereas Kx is fixated on x. But since K(Kx) and Kx are the same bird, then the same bird is fixated on both Kx and x, which makes Kx = x, according to the last problem. Therefore K is fond of x.

19 • We will show that the only way a kestrel can be egocentric is that it is the only bird in the forest!

Proof 1: Suppose that K is an egocentric kestrel. Then K is hopelessly egocentric, according to Problem 11. Now let x and y be any birds in the forest, and we will show that x = y.

Since K is hopelessly egocentric, then Kx = K and Ky = K, so Kx = Ky. Therefore, according to Problem 16, x = y. So any birds x and y in the forest are identical with each other, and there is only one bird in the forest. Since we are given that K is in the forest, then K is the only bird in the forest.

Proof 2: Again we use the fact that since K is egocentric, then K is hopelessly egocentric. Now let x be any bird in the forest. Then Kx is fixated on x, since K is a kestrel, and also Kx = K, since K is hopelessly egocentric. Therefore K is fixated on x, since Kx is fixated on x and Kx is the bird K. This proves that K is fixated on every bird x in the forest. But by Problem 17, K cannot be fixated on more than one bird, hence all the birds of the forest must be identical.

20 • Yes, it does. Suppose I is agreeable. Then for any bird x there is a bird y such that xy = Iy. But Iy = y, hence xy = y. Thus x is fond of y.

21 • Yes, it does. Suppose every bird x is fond of some bird y. Then xy = y, but also Iy = y, and so bird I agrees with x on the bird y.

22 • Both conclusions follow.

1. We are given that I is an identity bird and that any two birds are compatible. Now, take any bird B. Then B is compatible with I, so there are birds x and y such that Bx = y and Iy = x. Since Iy = x, then y = x, because y = Iy. Since y = x and Bx = y, then Bx = x, so B is fond of the bird x. Therefore every bird B is fond of some bird x.

2. This follows from the first conclusion and Problem 21.

23 • Suppose I is an identity bird and I is hopelessly egocentric. Take any bird x. Then Ix = I, since I is hopelessly egocentric, but also Ix = x, since I is an identity bird. Then x = I, so again we have the sad fact that there is only one bird in the forest; every bird x is identical with I.

24 • Suppose L is a lark and I is an identity bird. Then for any bird x, (LI)x = I(xx) = xx. Therefore LI is a mockingbird. This means that if someone calls out I to L, then L names a mockingbird.

25 • This is quite simple. Suppose L is a lark. Then for any birds x and y, (Lx)y = x(yy). This is also true when y is the bird Lx, and so (Lx)(Lx) = x((Lx)(Lx)). And so, of course, x((Lx)(Lx)) = (Lx)(Lx), which means that x is fond of the bird (Lx)(Lx). So every bird x is normal.

For help in solving future problems, we make a note of the fact that for any lark L, any bird x is fond of the bird (Lx)(Lx).

26 • We will show that if L is a hopelessly egocentric lark, then every bird is fond of L.

Suppose L is a lark and that L is hopelessly egocentric. Since L is hopelessly egocentric, then for any birds x and y, (Lx)y = L, according to Problem 14. In particular, taking Lx for y, (Lx)(Lx) = L. But x is fond of (Lx)(Lx), as we proved in the last problem. Therefore x is fond of L, since (Lx)(Lx) = L. This proves that every bird x is fond of L.

27 • This is an interesting proof! We have already proved in Problem 18 that if K is fond of Kx, then K is fond of x. In particular, taking K for x, if K is fond of KK, then K is fond of K.

Now suppose L is a lark of the forest and K is a kestrel of the forest and that L is fond of K. Then LK = K, hence (LK)K = KK. But (LK)K = K(KK), since L is a lark. Therefore KK = K(KK)—they are both equal to (LK)K—which makes K fond of KK. Then K is fond of K, as we showed in the last paragraph. Hence K is egocentric. Then by Problem 19, K is the only bird in the forest. But this contradicts the given fact that L is in the forest and L $\neq$ K.

28 • Suppose K is fond of L. Then by the solution to Problem 9, L is hopelessly egocentric. Therefore, by Problem 26, every bird is fond of L.

29 • Suppose the forest contains a lark L. Then by Problem 25, every bird is fond of at least one bird. In particular, the bird LL is fond of some bird y. (This constitutes our first trick!) Therefore (LL)y = y, but (LL)y = L(yy), because L is a lark, and so for *any* bird x, (Lx)y = x(yy). Therefore L(yy) = y, since they are both equal to (LL)y. Therefore (L(yy))y = yy. (This is our second trick!) But (L(yy))y = (yy)(yy). This can be seen by substituting (yy) for x in the equation (Lx)y = x(yy). So yy and (yy)(yy) are both equal to (L(yy))y, hence (yy)(yy) = yy, which means that yy is egocentric.

This proves that if y is any bird of whom LL is fond, then yy must be egocentric. Furthermore, LL *is* fond of some bird y, according to Problem 25.

We can actually compute a bird y of which LL is fond. We saw in the solution to Problem 25 that for any bird x, x is fond of (Lx)(Lx). Therefore LL is fond of (L(LL))(L(LL)). So we can take (L(LL))(L(LL)) for the bird y. Our egocentric bird is then ((L(LL))(L(LL)))((L(LL))(L(LL))).

CHAPTER 10

Is There a Sage Bird?

Inspector Craig of Scotland Yard was a man of many interests. His activities in crime detection, law, logic, number machines, retrograde analysis, vampirism, philosophy, and theology are familiar to readers of my earlier puzzle books. He was equally interested in ornithological logic—a field that applies combinatory logic to the study of birds. He was therefore delighted to hear about the bird forest of the last chapter and decided to visit it and do some "inspecting."

When he arrived, the first thing he did was to interview the bird sociologist of the forest, whose name was Professor Fowler. Professor Fowler told Craig of the two laws C_1 and C_2, the basic composition law and the existence of a mockingbird, from the first problem of the last chapter. From this, Inspector Craig was of course able to deduce that every bird was fond of at least one bird.

"However," explained Craig to Fowler, "I would like to go a bit more deeply into the matter. I am what mathematical logicians call a constructivist. I am not satisfied to know merely that given any bird x, there exists *somewhere* in the forest a bird y of which x is fond; I would like to know how, given a bird x, I can *find* such a bird y. Is there by any chance a bird in this forest that can supply such information?"

"I really don't understand your question," replied Fowler. "What do you mean by a bird's *supplying* such information?"

"What I want to know," said Craig, "is whether or not there is some special bird which, whenever I call out the name of a bird x to it, will respond by naming a bird of which x is fond. Do you know whether there is such a bird?"

"Oh, now I understand what you mean," said Fowler, "and your question is a very interesting one! All I can tell you is that it has been *rumored* that there is such a bird, but its existence in this forest has not been substantiated. Such birds are called *sage birds*—sometimes *oracle birds*—but, as I said, we don't know if there are any sage birds here. According to some history books, whose authenticity, however, is uncertain, sage birds were first observed in Greece—in Delphi, in fact—which might account for their also being called oracle birds. Accordingly, the Greek letter Θ is used to denote a sage bird. If there really is such a bird, then it has the remarkable property that for any bird x, x is fond of the bird Θx—in other words, $x(\Theta x) = \Theta x$. Or, as you might put it, if you call out x to Θ, then Θ will name a bird of which x is fond.

"I have been trying to find a sage bird for a long time now, but I'm afraid I haven't been very successful. If you could throw any light on the matter, I would be enormously grateful!"

Inspector Craig rose, thanked Professor Fowler, and told him that he would devote some thought to the matter. Craig then spent the day walking through the forest concentrating deeply on the problem. The next morning he returned to Professor Fowler.

"I doubt very much," said Craig, "that—from just the two conditions C_1 and C_2 that you have told me—it can be determined whether or not this forest contains a sage bird.

"The trouble is this," he explained: "We know that there is a mockingbird M. And we know that for any bird x there is *some* bird y that composes x with the mockingbird M. Then, as you know, x is fond of the bird yy. But given the bird x, how does one *find* a bird y that composes x with M? If there were some bird A that supplied this information, then the problem would be solvable. But from what you have told me, I have no reason to believe that there is such a bird."

"Oh, but there *is* such a bird," replied Fowler. "I'm sorry, but I forgot to tell you that we do have a bird A such that whatever bird x you call out to A, A will respond by naming a bird that composes x with M. That is, for any bird x, the bird Ax composes x with M."

"Splendid!" said Craig. 'That completely solves your problem: This forest *does* contain a sage bird."

How did Craig know this?

"Wonderful!" said Fowler, after Craig proved that the forest contained a sage bird.

"And now, what are your plans? You know, perhaps, that this forest is only one of a whole chain of remarkable bird forests. You should definitely visit Curry's Forest, and before you come to that, you will pass through a forest unusually rich in bird life. You will probably want to spend a good deal of time there; there is so much to learn!"

Craig thanked Professor Fowler and departed for the next forest. He little realized that this was only the beginning of a summer-long venture!

SOLUTION

This problem, though important, is really quite simple!

To begin with, the bird A described by Fowler is nothing more nor less than a lark! The reason is this: To say that for every bird x, the bird Ax composes x with M is to say that for any bird x and any bird y, $(Ax)y = x(My)$. But $My = yy$, so $x(My) = x(yy)$. Therefore the bird A described by Fowler satisfies the condition that for any birds x and y, $(Ax)y = x(yy)$, which means that A is a lark.

And so the problem boils down to this: Given a mockingbird M, a lark L, and the basic composition condition C_1, prove that the forest contains a sage bird.

Well, we have shown in the solution to Problem 25 of the last chapter that any bird x is fond of the bird $(Lx)(Lx)$, hence x is fond of $M(Lx)$, since $M(Lx) = (Lx)(Lx)$. Now, by the basic composition condition C_1, there is a bird Θ that composes M with L. This means that for any bird x, $\Theta x =$

M(Lx). Since x is fond of M(Lx) and M(Lx) = Θx, then x is fond of Θx, which means that Θ is a sage bird.

In short, any bird that composes M with L is a sage bird.

The theory of sage birds (technically *called fixed point combinators*) is a fascinating and basic part of combinatory logic; we have only scratched the surface. We will go more deeply into the theory of sage birds in a later chapter, but we must first turn our attention to some of the more basic birds, which we will do in the next two chapters.

CHAPTER 11

BIRDS GALORE

In the next bird forest Craig visited, the resident bird sociologist was named Professor Adriano Bravura. Professor Bravura had an aristocratic, somewhat proud bearing, which many mistook for haughtiness. Craig soon realized that this impression was quite misleading; Professor Bravura was an extremely dedicated scholar who, like many scholars, was often absentminded and abstracted, and this "abstractedness" was what was so often mistaken for detachment and lack of concern for other human beings. Actually, Professor Bravura was a very warmhearted person who took a great interest in his students. Craig learned an enormous amount from him—as will the reader!

"We have many, many interesting birds in this forest," said Bravura to Craig at the first interview, "but before I tell you about them, it will be best for me to explain to you a well-known abbreviation concerning parentheses."

Professor Bravura then took a pencil and a pad of paper and placed it so that Craig could see what he was writing.

"Suppose I write down xyz," said Bravura. "Without further explanation, this notation is ambiguous; you cannot know whether I mean (xy)z or x(yz). Well, the convention is that we will mean (xy)z—or, as we say, if parentheses are omitted, they are to be restored *to the left*. This is the tradition in combinatory logic, and after a little practice, it makes complex expressions more easily readable.

"The same convention applies to even more complex expressions—for example, let us look at the expression (xy)zw. We look at (xy) as a unit, and so (xy)zw is really ((xy)z)w. What is xyzw? We first restore parentheses to the leftmost part, which is xy, and so xyzw is (xy)zw, which in turn is ((xy)z)w. And so xyzw is simply an abbreviation for ((xy)z)w.

"Other examples," said Bravura: "x(yz)w = (x(yz))w, whereas x(yzw) = x((yz)w).

"I think you should now try the following exercises to be sure that you fully grasp the principle of restoring parentheses to the left."

Here are the exercises Bravura gave Craig; the answers are given immediately afterward.

Exercises: In each of the following cases, fully restore parentheses to the left.

a. xy(zwy)v = ?
b. (xyz)(wvx) = ?
c. xy(zwv)(xz) = ?
d. xy(zwv)xz = ? *Note:* The answer is different from that for (c)!
e. x(y(zwv))xz = ?
f. Is the following true or false?

$$xyz(AB) = (xyz)(AB)$$

g. Suppose $A_1 = A_2$. Can we conclude that $BA_1 = BA_2$? And can we conclude that $A_1B = A_2B$?

h. Suppose xy = z. Which of the following conclusions is valid? *Note:* Tricky and important!

    1. xyw = zw
    2. wxy = wz

Answers:

a. xy(zwy)v = ((xy)((zw)y))v
b. (xyz)(wvx) = ((xy)z)((wv)x)
c. xy(zwv)(xz) = ((xy)((zw)v))(xz)

d. xy(zwv)xz = (((xy)((zw)v))x)z

e. x(y(zwv))xz = ((x(y((zw)v)))x)z

f. True; both sides reduce to ((xy)z)(AB).

g. Both conclusions are correct.

h. Suppose xy = z.

1. xyw = zw says that (xy)w = zw, and this is correct, since the birds (xy) and xy are identical and we are given that xy = z, and hence (xy) = z.

2. wxy = wz says that (wx)y = wz, and this certainly does *not* follow from the fact that xy = z! What *does* follow is that w(xy) = wz, but this is very different from (wx)y = wz.

So the first conclusion follows, but the second does not.

BLUEBIRDS

"Now that we have gone through these preliminaries," said Bravura, "we can get on to the more interesting things about this forest.

"As I have told you, we have many fascinating birds here. A bird of basic importance is the *bluebird*—by which I mean a bird B such that for all birds x, y, z, the following holds:

$$Bxyz = x(yz)$$

"In unabbreviated notation," said Bravura, "I would have written: ((Bx)y)z = x(yz). However, I find it much easier to read: Bxyz = x(yz)."

· I ·

"Why are bluebirds of basic importance?" asked Craig.

"For many reasons, which you will see," replied Bravura. "For one thing, if a forest contains a bluebird—which this forest fortunately does—then the basic composition law must hold: For any bird C and D, there is a bird E that composes C with D. Can you see why?"

Note: Recall from Chapter 8 that if E composes C with D, it means that for every bird x, Ex = C(Dx).

2 · BLUEBIRDS AND MOCKINGBIRDS

"Suppose," said Bravura, "that a bird forest contains a bluebird B and a mockingbird M. Since B is present, the composition law holds, as you have just seen. Therefore, as you know, it follows that every bird x is fond of some bird. However, since B is present, you can write down an expression in terms of B, M, x that describes a bird of which x is fond. Can you see how to write down such an expression?"

3 · EGOCENTRIC

"Given a bluebird B and a mockingbird M," said Bravura, "can you see how to write down an expression for an egocentric bird?"

4 · HOPELESSLY EGOCENTRIC

"Now," said Bravura, "suppose a forest contains a bluebird B, a mockingbird M, and a kestrel K. See if you can write down an expression in terms of B, M, and K for a hopelessly egocentric bird."

SOME DERIVATIVES OF THE BLUEBIRD

"And now," said Bravura, "let us forget about mockingbirds and kestrels for a while and concentrate on just the bluebird B. From just this one bird alone, many useful birds can be derived. Not all of them are of major importance, but several of them will crop up from time to time in the course of your study."

5 · DOVES

"For example, one fairly important bird is the *dove,* by which is meant a bird D such that for any birds x, y, z, w, the following condition holds:

$$Dxyzw = xy(zw)$$

"The bird D can be derived from B alone. Can you see how?"

6 · BLACKBIRDS

"Then," said Bravura, "there is the *blackbird*—a bird B_1 such that for any birds x, y, z, w, the following condition holds:

$$B_1xyzw = x(yzw)$$

"Prove that any forest containing a bluebird must also contain a blackbird.

"Of course," added Bravura, "in deriving a blackbird from a bluebird, you are free to use the dove D if that is helpful, since you have already seen how D can be derived from B."

7 · EAGLES

"Then there is the *eagle*," said Bravura, "by which is meant a bird E such that for any birds x, y, z, w, v, the following condition holds:

$$Exyzwv = xy(zwv)$$

"The eagle can be derived from just the bird B. Can you see how? Again, it will simplify your derivation to use birds already derived from B."

8 · BUNTINGS

"A *bunting*," said Bravura, "is a bird B_2 satisfying the following condition—for any birds x, y, z, w, v, of course:

$$B_2xyzwv = x(yzwv)$$

"Given B, find a bunting B_2."

9 · DICKCISSELS

Bravura continued: "By a *dickcissel* I mean a bird D_1 satisfying the following condition:

$$D_1xyzwv = xyz(wv)$$

"Show how a dickcissel D_1 can be derived from a bluebird B."

10 · BECARDS

"Then there is the *becard*," said Bravura, "a bird B_3 such that for all birds x, y, z, w, the following condition holds:

$$B_3xyzw = x(y(zw))$$

"Can you see how to derive a becard from a bluebird, and from any other birds already derived from B?"

11 · DOVEKIES

"Then there is the *dovekie,*" said Bravura, "which is a bird D_2 satisfying the following condition:

$$D_2xyzwv = x(yz)(wv)$$

"Can you see how to derive a dovekie D_2 from a bluebird B?"

12 · BALD EAGLES

"And now," said Bravura, "given a bluebird B, see if you can derive a *bald eagle*—a bird $\hat{E}$ such that for all birds x, y_1 y_2, y_3, z_1, z_2, z_3, the following condition holds:

$$\hat{E}xy_1\, y_2, y_3, z_1, z_2, z_3 = x(y_1\, y_2, y_3)\, (z_1, z_2, z_3).$$

"I think you have had enough problems for today," said Bravura. "We have now derived eight different birds from the one bird B. We could derive many more, but I think you have seen enough to get a good feeling for the behavior of the bluebird. All these birds—including B—belong to a family of birds known as *compositors*. They serve to introduce parentheses. The only two that you need remember are the bluebird B and the dove D; they are standard in the literature of combinatory logic. The other seven birds don't have standard names, but I have found it convenient to give them names, as some of them will crop up again.

"Tomorrow, I will tell you about some very different birds."

SOME OTHER BIRDS

Inspector Craig returned bright and early the next morning. He was surprised to find Professor Bravura in the garden, seated at a table with paper, pencils, and piles of notes. Two cups of freshly brewed steaming hot coffee had been laid out.

13 · WARBLERS

"Too beautiful a morning to work indoors," said Bravura. "Besides, I may be able to show you some of the birds we discuss.

"Ah, there goes a warbler!" said Bravura. "This bird W is an important bird and is quite standard in combinatory logic. It is defined by the following condition:

$$Wxy = xyy$$

"Do not confuse this with the lark L!" cautioned Bravura. "Remember, $Lxy = x(yy)$, whereas $Wxy = xyy$. These are very different birds!

"I have a nice little problem for you," continued Bravura. "Prove that any forest containing a warbler W and a kestrel K must contain a mockingbird M."

After a bit of time, Bravura said, "I see you are having difficulty. I think I will first give you two simpler problems."

· 14 ·

"Show that from a warbler W and an identity bird I we can get a mockingbird."

Craig solved this quite easily.

· 15 ·

"Now show that from a warbler W and a kestrel K we can get an identity bird."

"Oh, I get the idea!" said Craig.

16 · CARDINALS

Just then, a brilliant red bird flew by.

"A cardinal!" said Bravura. "One of my favorite birds! It also plays a basic role in combinatory logic. The cardinal C is defined by the following condition:

$$Cxyz = xzy$$

"The cardinal belongs to an important family of birds known as *permuting* birds. You see that in the above equation, the variables y and z have got switched around.

"Here's an easy problem for you," said Bravura. "Prove that any forest containing a cardinal and a kestrel must contain an identity bird."

17 · THRUSHES

"A bird closely related to the cardinal is the thrush," said Bravura. "Why, there is one right over there! A thrush T is defined by the following condition:

$$Txy = yx$$

"The thrush is the simplest of the permuting birds," said Bravura. "It is derivable from a cardinal C and an identity bird I. Can you see how?"

18 · COMMUTING BIRDS

"Two birds x and y are said to commute," said Bravura, "if xy = yx. This means that it makes no difference whether you call out y to x, or x to y; you get the same response in either case.

"There's an interesting thing about thrushes," said Bravura. "If a forest contains a thrush, and if every bird of the forest is fond of some bird, then there must be at least one bird A that commutes with every bird. Can you see how to prove this?"

· 19 ·

"Given a bluebird B, a thrush T, and a mockingbird M," said Bravura, "find a bird that commutes with every bird."

BLUEBIRDS AND THRUSHES

"Bluebirds and thrushes work beautifully together!" said Bravura. "From these two birds, you can derive a whole variety of birds known as *permuting* birds. For one thing, from a bluebird B and a thrush T, you can derive a cardinal—this was discovered by the logician Alonzo Church in 1941."

"That sounds interesting," said Craig. "How is it done?"

"The construction is a bit tricky," said Bravura. "Church's expression for a cardinal C in terms of B and T has eight letters, and I doubt that it can be done with fewer. I will simplify the problem for you by first deriving another bird—one useful in its own right."

20 · ROBINS

"From B and T," said Bravura, "we can derive a bird R called a *robin* which satisfies the following condition:

$$Rxyz = yzx$$

"Given a bluebird and a thrush, do you see how to derive a robin?"

21 · ROBINS AND CARDINALS

"And now, from just the robin alone, we can derive a cardinal. Can you see how? The solution is quite pretty!"

A bonus question: "Putting the last two problems together," said Bravura, "you can see how to derive C from B and T. However, the solution you then get will contain nine letters. It can be shortened by one letter. Can you see how?"

22 · TWO USEFUL LAWS

"The following two laws are useful," said Bravura. "We let R be BBT and C be RRR. Prove that for any bird x, the following facts hold:

a. $Cx = RxR$
b. $Cx = B(Tx)R$."

23 · A QUESTION

"You have just seen that a cardinal can be derived from a robin. Can a robin be derived from a cardinal?"

24 · FINCHES

"Ah, there goes a finch!" said Bravura. "A finch is a bird F satisfying the following condition:

$$Fxyz = zyx$$

"The finch is another permuting bird, of course, and it can also be derived from B and T. This can be done in several ways. For one thing, a finch can be easily derived from a bluebird, a robin, and a cardinal—and hence from a bluebird and a robin or from a bluebird and a cardinal. Can you see how?"

· 25 ·

"Alternatively, a finch can be derived from a thrush T and an eagle E. Can you see how?"

· 26 ·

"Now you have available two methods of expressing a finch in terms of a bluebird B and a thrush T. You will see that one of them yields a much shorter expression than the other."

27 · VIREOS

"Ah, there goes a vireo!" said Bravura in some excitement. "If you ever get to study combinatorial birds in relation to arithmetic—as doubtless you will—you will find the vireo to be of basic importance. The vireo V is also a permuting bird—it is defined by the following condition:

$$Vxyz = zxy$$

"The vireo has a sort of opposite effect to the robin," commented Bravura. "This bird is also derivable from B and T. One way is to derive it from a cardinal and a finch. Can you see how?"

· 28 ·

"How would you most easily express a vireo in terms of a finch and a robin?" asked Bravura. "It can be done with an expression of only three letters."

29 · A QUESTION

"I will later show you another way of deriving a vireo," said Bravura. "Meanwhile I'd like to ask you a question. You have seen that a vireo is derivable from a cardinal and a finch. Is a finch derivable from a cardinal and a vireo?"

30 · A CURIOSITY

"Another curiosity," said Bravura. "Show that any forest containing a robin and a kestrel must contain an identity bird."

SOME RELATIVES

It was now about noon, and Mrs. Bravura—an exceedingly beautiful, delicate, and refined Venetian lady—brought out a magnificent lunch. After the royal repast, the lesson continued. "I should now like to tell you about some useful relatives of the cardinal, robin, finch, and vireo," said Bravura. "All of them can be derived from just the two birds B and T—in fact, from B and C."

31 · THE BIRD C*

"First there is the bird C* called a *cardinal once removed*, satisfying the following condition:

$$C^*xyzw = xywz$$

"Notice," said Bravura, "that in this equation, if we erased x from both sides, and also erased the star, we would have the true statement Cyzw = ywz."

"This is the idea behind the term 'once removed.' The bird C* is like C, except that its action is 'deferred' until we skip over x; we then 'act' on the expression yzw as if we were using a cardinal."

"And now see if you can derive C*from B and C. This is quite simple!"

32 · THE BIRD R*

"The bird R*—a *robin once removed*—bears much the same relation to R as C* does to C. It is defined by the following condition:

$$R^*xyzw = xzwy$$

"Show that R* is derivable from B and C—and hence from B and T."

33 · THE BIRD F*

"By *a finch once removed* we mean a bird F* satisfying the following condition:

$$F^*xyzw = xwzy$$

"Now derive F* from birds derivable from B and C."

34 · THE BIRD V*

"Finally, we have the *vireo once removed*—a bird V* satisfying the following condition:

$$V^*xyzw = xwzy$$

"Show how to derive V* from birds derivable from B and C."

35 · TWICE REMOVED

"Given birds B and C, find birds C**, R**, F**, V** such that for any birds x, y, z_1, z_2, z_3 the following conditions hold:

$$C^{**}xyz_1z_2z_3 = xyz_1z_3z_2$$
$$R^{**}xyz_1z_2z_3 = xyz_2z_3z_1$$
$$F^{**}xyz_1z_2z_3 = xyz_3z_2z_1$$
$$V^{**}xyz_1z_2z_3 = xyz_3z_1z_2$$

"These are the birds C, R, F, V *twice* removed. They will occasionally be useful."

36 · VIREOS REVISITED

"You have seen that a vireo is derivable from a cardinal and a finch. It is also derivable from the two birds C* and T. Can you see how?"

QUEER BIRDS

"And now," said Bravura, "we turn to an interesting family of birds which both parenthesize and permute. They are all derivable from B and T."

37 · QUEER BIRDS

"The most important member of the family is the *queer* bird Q defined by the following condition:

$$Qxyz = y(xz)$$

"As you can see, Q both introduces parentheses and permutes the order of the letters x and y.

"A comparison of Q with the bluebird B is worth noting: For any birds x and y, the bird Bxy composes x with y, whereas Qxy composes y with x.

"The bird Q is quite easily derived from B and one other bird that you have already derived from B and T. Can you see which one and how?"

38 · QUIXOTIC BIRDS

"The queer bird Q has several cousins; perhaps the most important one is the *quixotic* bird Q_1 defined by the condition:

$$Q_1xyz = x(zy)$$

"Show that Q_1 is derivable from B and T. Again, you may of course use any birds previously derived from B and T."

39 · QUIZZICAL BIRDS

"Then there is the *quizzical* bird Q_2—another cousin of Q. It is defined by the condition:

$$Q_2xyz = y(zx)$$

"Show that Q_2 is derivable from B and T."

40 · A PROBLEM

"Here is a little problem for you," said Bravura. "Suppose we are given that a certain bird forest contains a cardinal, but we are not given that it contains a bluebird or a thrush. Prove that if the forest contains either a quixotic bird or a quizzical bird, then it must contain the other as well."

41 · QUIRKY BIRDS

"Another cousin of Q is the *quirky* bird Q_3 defined by the following condition:

$$Q_3xyz = z(xy)$$

"Show that Q_3 is derivable from B and T."

42 · QUACKY BIRDS

"The last cousin of Q is the *quacky* bird Q_4 defined by the following condition:

$$Q_4xyz = z(yx)."$$

"What a strange name!" exclaimed Craig.

"I didn't name it; it was named after a certain Professor Quack, who discovered it. Anyhow, can you see how to derive it from B and T?"

43 · An Old Proverb

"There is an old proverb," said Bravura, "that says that if a cardinal is present, then you can't have a quirky bird without a quacky bird, or a quacky bird without a quirky bird. And if there isn't such a proverb, then there should be! Can you see why the proverb is true?"

44 · A Question

"Is a quacky bird derivable from Q_1 and T?"

45 · An Interesting Fact About the Queer Bird Q

"You have seen that the queer bird Q is derivable from the bluebird B and the thrush T. It is of interest that you can alternatively derive a bluebird B from a queer bird Q and a thrush T. Can you see how? The method is a bit tricky!"

· 46 ·

"One can derive a cardinal C from Q and T more easily than from B and T—in fact, you need an expression of only four letters. Can you find it?"

47 · Goldfinches

"Another bird derivable from B and T which I have found useful is the *goldfinch* G defined by the following condition:

$$Gxyzw = xw(yz)$$

"Can you see how to derive it from B and T?"

"We could go on endlessly deriving birds from B and T," said Bravura, "but it is now getting chilly and Mrs. Bravura has prepared a nice dinner for us. Tomorrow I will tell you about some other birds."

SOLUTIONS

1 • Given a bluebird B, we are to show that for any birds C and D, there is a bird E that composes C with D. Well, BCD is such a bird E, because for any bird x, (BCD)x = ((BC)D)x = C(Dx). Therefore BCD composes C with D.

2 • We saw in the solution to Problem 1 of Chapter 9 that if y is any bird that composes x with M, then x is fond of the bird yy. Now, BxM composes x with M (according to the last problem), and so x must be fond of (BxM)(BxM).

Let us double-check: (BxM)(BxM) = BxM(BxM) = x(M(BxM)) = x((BxM)(BxM))—because M(BxM) = (BxM)(BxM). So (BxM)(BxM) = x((BxM)(BxM)), or what is the same thing, x((BxM)(BxM)) = (BxM)(BxM), which means that x is fond of the bird (BxM)(BxM).

The expression (BxM)(BxM) can be shortened to M(BxM). So x is fond of M(BxM).

3 • We have just seen that for any bird x, x is fond of M(BxM). If we take x to be the mockingbird M, then M is fond of M(BMM). Now, in the solution to Problem 2 in Chapter 9, we saw that any bird of which the mockingbird is fond must be egocentric. Therefore M(BMM) is egocentric.

Let us double-check: M(BMM) = (BMM)(BMM) = BMM(BMM) = M(M(BMM)) = (M(BMM))(M(BMM)). And so we see that M(BMM) = (M(BMM))(M(BMM)), or what is the same thing, (M(BMM))(M(BMM)) = M(BMM), which means that M(BMM) is egocentric.

4 • Since for any bird x, x is fond of M(BxM), then the kestrel K is fond of M(BKM). Therefore M(BKM) is hopelessly egocentric, according to the solution of Problem 9 of Chapter 9, in which we saw that any bird of which the kestrel is fond must be hopelessly egocentric.

5 • It is sometimes easiest to work these problems backward. We are look-ing for a bird D such that Dxyzw = (xy)(zw). Let us look at the expression (xy)(zw) and see how we can get back to Dxyzw, where D is the bird to be found. Well, we look at the expression (xy) as a unit—call it A—and so (xy)(zw) = A(zw), which we recognize as BAzw, which is B(xy)zw. So the first step of the "backward" argument is to recognize (xy)(zw) as B(xy)zw. Next, we look at the front end B(xy) of the expression and rec-ognize it as BBxy. And so B(xy)zw is BBxyzw. Therefore we take D to be the bird BB.

Let us double-check by running the argument forward.

Dxyzw = BBxyzw, since D = BB.

$\qquad$ = B(xy)zw, since BBxy = B(xy).

$\qquad$ = (xy)(zw) = xy(zw)

6 • Since we have already found the dove D from B, we are free to use it. In other words, in any solution for B_1 in terms of B and D, we can replace D by BB, thus getting a solution in terms of B alone.

Again we will work the problem backward.

x(yzw) = x((yz)w) = Bx(yz)w. We recognize Bx(yz) as DBxyz, and so Bx(yz)w = DBxyzw. Therefore x(yzw) = DBxyzw, or what is the same thing, DBxyzw = x(yzw). We can therefore take B_1 to be the bird DB. The reader can check the solution by running the argument forward.

In terms of B alone, B_1 = (BB)B, which also can be written B_1 = BBB.

7 • We will use the bird B_1 found in the last problem. Again we will work the problem backward.

xy(zwv) = (xy)(zwv). Looking at (xy) as a unit, we can see that (xy)(zwv) = B_1(xy)zwv. Also B_1(xy) = BB_1xy, so B_1(xy)zwv = BB_1yzwv. And so we take E to be the bird BB_1.

In terms of B alone, E = BB_1 = B(BBB).

To illustrate a point, suppose we tried to find E directly from B, with-out using any birds previously derived from B. We could proceed as fol-lows:

We look at the expression xy(zwv). The first thing we try to do is to free the last letter v from parentheses. Well, xy(zwv) = (xy)((zw)v) =

B(xy)(zw)v. Now we have freed v from parentheses. We next work on the expression B(xy)(zw), and we would like to free w from parentheses. Looking at B(xy) as a unit, we see that B(xy)(zw) = B(B(xy))zw. We have now freed w from parentheses, and as good fortune would have it we have freed z as well. It remains merely to work on B(B(xy)). We wish to free y from parentheses, but since it is enclosed in two pairs of parentheses, we first free it from the outer pair. Well, B(B(xy)) = BBB(xy). We now look at BBB as a unit and see that BBB(xy) = B(BBB)xy. And so we take E to be B(BBB), which is the same solution we got before.

In this analysis, we have substantially duplicated the labor of deriving the bird B_1, and had this problem been posed *before* Problem 6, we would have had to do this. The moral is that in solving these problems, the reader should be on the lookout for solutions to earlier problems that might be helpful.

8 • Starting from scratch, the solution would be long. Using the eagle of the last problem, the solution is easy:

$$x(yzwv) = x((yzw)v) = Bx(yzw)v$$

But Bx(yzw) is EBxyzw, so Bx(yzw)v = EBxyzwv. So we take B_2 to be EB.

In terms of B alone, B_2 = B(BBB)B.

9 • There are two ways we can go about this which will be interesting to compare.

Our first method uses the dove D. Now, xyz(wv) = (xy)z(wv). Looking at (xy) as a unit, we see that (xy)z(wv) = D(xy)zwv. Also D(xy) = BDxy, and so D(xy)zwv = BDxyzwv. And so we take D_1 to be BD, which in terms of B alone is B(BB).

We can also look at the matter this way: xyz(wv) = (xyz)(wv). Looking at (xyz) as a unit, we see that (xyz)(wv) = B(xyz)wv. However, B(xyz) we recognize as B_1Bxyz. Therefore B_1B is also a solution.

Now, B_1 = BBB, so B_1B = BBBB. But BBBB = B(BB), and so we really get the same solution.

10 • We use the bird D_1 of the last problem. Looking at (zw) as a unit, $x(y(zw)) = Bxy(zw) = D_1Bxyzw$. So we take B_3 to be D_1B.

In terms of B alone, $B_3 = B(BB)B$.

11 • Again, we can go about this two ways. On the one hand, if we look at (yz) as a unit, then $x(yz)(wv) = Dx(yz)wv$. Also $Dx(yz) = DDxyz$, and so we can take D_2 to be DD, which in terms of B is $BB(BB)$.

On the other hand, we can look at $x(yz)$ as a unit and see that $x(yz)(wv) = B(x(yz))wv$. But $B(x(yz)) = B_3Bxyz$, and so B_3B is also a solution.

It is really the same solution, since $B_3B = B(BB)BB = BDBB = D(BB)$ $= DD$, which in turn is $BB(BB)$.

We might remark that we have proved a stronger result than was called for: We were required to derive D_2 from B, but we have in fact succeeded in deriving it from D, since $D_2 = DD$. Therefore if we were not told that the forest contains a bluebird, but were given only the weaker condition that it contains a dove, this would still be enough to imply that the forest contains a dovekie.

12 • We will prove the stronger result that if the forest contains an eagle (without necessarily containing a bluebird) then it must contain a bald eagle.

Looking at $(y_1y_2y_3)$ as a unit, we see that $x(y_1y_2y_3)(z_1z_2z_3) = Ex(y_1y_2y_3)z_1z_2z_3$. But $Ex(y_1y_2y_3) = EExy_1y_2y_3$, and so $x(y_1y_2y_3)(z_1z_2z_3) = Ex(y_1y_2y_3)z_1z_2z_3 = EExy_1y_2y_3z_1z_2z_3$. And so we take E to be EE.

In terms of B, the bird EE is $B(BBB)(B(BBB))$.

13, 14, and 15 • First, we shall do Problem 14: Given W and I, the bird WI is a mockingbird, because for any bird x, $WIx = Ixx = xx$, since $Ix = x$.

Now for Problem 15: Given W and K, the bird WK is an identity bird, because for any bird x, $WKx = Kxx = x$.

Putting these two problems together, WK is an identity bird, and hence W(WK) should be a mockingbird by Problem 14. Let us check:

$$W(WK)x = WKxx = (WKx)x = (Kxx)x = xx.$$

Yes, $W(WK)$ is a mockingbird. This solves Problem 13.

16 • For any bird A whatsoever, the bird CKA is an identity bird, because for any bird x, $CKAx = KxA = x$. So, for example, CKK is an identity bird; so is CKC.

17 • CI is a thrush, because for any birds x and y, $CIxy = Iyx = yx$.

18 • The given condition of the problem implies that the thrush T is fond of some bird A. Thus $TA = A$. Then for any bird x, $TAx = Ax$. Also $TAx = xA$, since T is a thrush. Therefore $Ax = xA$, and so A commutes with every bird x.

19 • Given the bluebird B and the mockingbird M, as well as the thrush T, we know from Problem 2 of this chapter that T is fond of the bird M(BTM). Remember that for *any* bird x, x is fond of M(BxM). Therefore, according to the last problem, M(BTM) commutes with every bird.

20 • We will work the problem backward: $yzx = Tx(yz)$. We recall the dove D and we see that $Tx(yz) = DTxyz$. Therefore we take R to be DT. In terms of B and T alone, $R = BBT$.

21 • Working the problem backward, with only a robin available, we find the solution virtually forced on us! We want to get xzy back into the position xyz. Well, $xzy = Ryxz$—what else can we do? Now, $Ryx = RxRy$—again, what other move could we make? Finally, $RxR = RRRx$.

Retracing our steps, $RRRx = RxR$, hence $RRRxy = RxRy = Ryx$. Since $RRRxy = Ryx$, then $RRRxyz = Ryxz = xzy$. Therefore we take our cardinal C to be the bird RRR.

A bonus question: When written in terms of B and T, C = (BBT)(BBT)(BBT). This expression can be shortened by one letter: $C = RRR = BBTRR = B(TR)R$, since $BBTR = B(TR)$. So $C = B(T(BBT))(BBT)$.

The expression B(T(BBT))(BBT) has only eight letters and is Alonzo Church's expression for a cardinal. Personally, I find it easier to remember the cardinal as RRR.

22 • a.Cx = RRRx = RxR

b. Since Cx = RxR and R = BBT, then Cx = BBTxR = B(Tx)R.

23 • Yes; CC is a robin, because CCxy = Cyx, hence CCxyz = Cyxz = yzx.

24 • We will work the problem backwards: zyx = Rxzy = (Rx)zy = C(Rx)yz = BCRxyz. And so we take F to be BCR.

25 • We can also analyze the situation this way: zyx = Tx(zy) = Tx(Tyz) = ETxTyz = (ETx)Tyz = TT(ETx)yz, because (ETx)T = TT(ETx). Continuing, TT(ETx) = ETTETx, hence TT(ETx)yz = ETTETxyz. Therefore we can take F to be ETTET.

26 • If we take F to be BCR, as in Problem 24, then in terms of B and T, the bird F = B(B(T(BBT))(BBT))(BBT).

We get a shorter solution if we express F as ETTET and then reduce to B and T. This is done as follows: ETTET = B(BBB)TTET, because E = B(BBB). Now B(BBB)TTET = BBB(TT)ET = B(B(TT))ET = B(TT)(ET) = B(TT)(B(BBB)T). And so we get a solution shorter by four letters.

27 • zxy = Fyxz = CFxyz, because Fyx = CFxy. We therefore can take V to be CF.

28 • According to law (a) stated in Problem 22, CF = RFR, and CF is a vireo. So RFR is a vireo.

29 • Yes; CV is a finch, because CVxyz = Vyxz = zyx.

30 • For any bird A, the bird RAK must be an identity bird, because RAKx = KxA = x. So, for example, RRK and RKK are both identity birds.

31 • xywz = (xy)wz = C(xy)zw = BCxyzw. And so we take C^* to be BC.

32 • Actually, we can get the bird R^* from just C^*: xzwy = C^*xzyw. Also C^*xzy = C^*C^*xyz, therefore C^*xzyw = C^*C^*xyzw. So, xzwy = C^*C^*xyzw. We therefore take R^* to be C^*C^*.

33 • We can get F^* from B, C^*, and R^* as follows: xwzy = R^*xywz = (R^*x)ywz = $C^*(R^*x)$yzw = BCR^*xyzw, since $C^*(R^*x)$ = BCR^*x, so we take F^* to be BCR^*.

34 • Just as we got V from C and F (V = CF), we can get V^* from C^* and F^*.

xwyz = F^*xzyw = C^*F^*xyzw, because F^*xzy = C^*F^*xyz. And so we take V^* to be C^*F^*.

35 • The secret here is remarkably simple! Take C^{**} to be BC*; R^{**} to be BR*; F^{**} to be BF*; and V^{**} to be BV*.

36 • C^*T is a vireo, because C^*Txyz = Txzy = zxy. This means that BCT is a vireo.

37 • We can get Q from a bluebird B and a cardinal C as follows:

y(xz) = Byxz = CBxyz, since Byx = CBxy.

And so we take Q to be CB.

In terms of B and T, Q = CB = RRRB = RBR = BBBTBR = B(TB)R = B(TB)(BBT).

38 • We will now find a good use for the starred birds of "some relatives of bluebirds and thrushes." x(zy) = Bxzy = C^*Bxyz. We can therefore take Q_1 to be C^*B. In terms of B and C, we take Q_1 to be BCB.

39 • y(zx) = Byzx = R^*Bxyz. We can therefore take Q_2 to be R^*B. In terms of B and C, we take Q_2 to be BC(BC)B or, more simply, C(BCB).

40 • Suppose the forest contains a cardinal C. If a quixotic bird Q_1 is present, a CQ_1 must be a quizzical bird, because $CQ_1xyz = Q_1yxz = y(zx)$. On the other hand, if a quizzical bird Q_2 is present, then CQ_2 must be a quixotic bird, because $CQ_2xyz = Q_2yxz = x(zy)$.

41 • $z(xy) = Bzxy = V^*Bxyz$. We can therefore take Q_3 to be V^*B.

However, Q_3 can be gotten directly from B and T much more simply: $z(xy) = T(xy)z = BTxyz$. And so it is simpler to take Q_3 to be BT.

42 • $z(yx) = Bzyx = F^*Bxyz$. And so we can take Q_4 to be F^*B. Another solution follows from the next problem.

43 • Suppose a cardinal C is present. If a quirky bird Q_3 is present, then CQ_3 must be a quacky bird, because $CQ_3xyz = Q_3yxz = z(yx)$. On the other hand, if a quacky bird Q_4 is present, then CQ_4 must be a quirky bird, because $CQ_4xyz = Q_4yxz = z(xy)$.

Since BT is a quirky bird, then C(BT) is a quacky bird, and so for Q_4 we can take C(BT) instead of F^*B.

44 • Yes; Q_1T is a quacky bird, since $Q_1Txyz = T(yx)z = z(yx)$.

Since we can take Q_1 to be BCB, then $Q_1T = BCBT = C(BT)$, and we get the same solution as if we took Q_4 to be CQ_3.

45 • $QT(QQ)$ is a bluebird because $QT(QQ)xyz = QQ(Tx)yz = Tx(Qy)z = Qyxz = x(yz)$.

46 • $QQ(QT)$ is a cardinal, since $QQ(QT)xyz = QT(Qx)yz = Qx(Ty)z = Ty(xz) = (xz)y = xzy$.

47 • $xw(yz) = Cx(yz)w = B(Cx)yzw = BBCxyzw$. And so we take G to be BBC.

The bird G has some curious properties, as we will see later on.

CHAPTER 12

MOCKINGBIRDS, WARBLERS, AND STARLINGS

MORE ON MOCKINGBIRDS

Inspector Craig returned early the next morning and again found Professor Bravura in the garden. The first thing that struck Craig was the singing of a distant bird whose song was the strangest that Craig had ever heard. It seemed totally disjointed; first there was a simple melodic line and then, out of the blue, a trill that seemed totally unrelated to the melody. Then followed a melody in a completely unrelated key!

"You've never heard a mockingbird before?" asked Bravura, who noticed Craig's astonishment.

"I guess not! It sounds almost mad!"

"Oh, well," said Bravura, "it remembers bits and snatches from the other birds and doesn't always put them together in the most logical order. I must say, though, that this particular mockingbird sounds wilder than any I've ever heard.

"Let me tell you some combinatorial properties of the mockingbird M," continued Bravura. "It has what is called a *duplicative* effect—it causes repetition of variables. It has this in common with the lark and the warbler. No bird derivable from B and T can have a duplicative effect, so the mockingbird is quite independent of them—it is definitely *not* derivable from B and T. But from the *three* birds B, T, and M, a whole variety of important birds can be derived."

I · THE BIRD M₂

"A very simple, but useful, example is the bird M_2—which I sometimes call a 'double' mockingbird—defined by the condition:

$$M_2xy = xy(xy)$$

"This bird is derivable from just B and M. That's pretty obvious, isn't it?"

2 · LARKS

"You recall the lark L satisfying the condition $Lxy = x(yy)$. Well, L is derivable from B, T, and M. One way is to derive it from B, C, and M, or from B, R, and M. Can you see how?"

· 3 ·

"I might mention, incidentally, that L is also derivable from the bluebird B and the warbler W. Can you see how? Actually, this fact is rather important."

· 4 ·

"My favorite construction of a lark," said Bravura, "uses just the mockingbird M and the queer bird Q. It is also the simplest! Can you see how it's done?"

WARBLERS

Just then a warbler flew by.

"Tell me," said Craig, "can a warbler be derived from B, T, and M? Since a lark can, I would not be too surprised if a warbler can."

"Ah, that's a good question," replied Bravura, "and it has a fascinating history. The logician Alonzo Church was interested in the entire class of birds derivable from the four birds B, T, M, and I. My forest happens to follow the thinking of Church; all my birds are derivable from B, T, M, and I. Now, in 1941, Church showed how to derive a warbler from B, T, M, and I. His method was both bizarre and ingenious; his expression for W in terms of B, T, M, and I involved twenty-four letters and thirteen pairs of parentheses! I will tell you about it another time." *Note to reader:* I discuss this in some of the exercises of this chapter.

"Shortly after," continued Bravura, "the logician J. Barkley Rosser found a much shorter expression—one with only ten letters. In looking at his expression, I noticed that he didn't use the identity bird I at all, hence your guess was correct: A warbler can be derived from just B, T, and M. It can be derived even more simply from B, C, and M—and more simply still from B, C, R, and M. But first let me tell you about another bird closely related to W."

5 · THE BIRD W′

"Show that from B, T, and M you can derive a bird W′ satisfying the following condition:

$$W'xy = yxx$$

"We might call W′ a *converse* warbler," said Bravura. "Curiously enough, W′ is easier to derive than W. It is particularly simple to derive W′ from B, R, and M. Can you see how?"

6 · THE WARBLER

"Now that you have W′, it is simple to get W. In fact, W can be derived from B, R., C, and M using an expression of only four letters. Can you see how?"

· 7 ·

"Now express W in terms of B, T, and M. This can be done with an expression of only ten letters, and there are two such expressions."

8 · A QUESTION

"You now see that W is derivable from B, T, and M. Is a mockingbird M derivable from B, T, and a warbler W?"

9 · TWO RELATIVES OF W

"We will occasionally have use for a bird W* satisfying the condition W*xyz = xyzz. How do you derive W* from B, T, and M? And what about a bird W** satisfying the condition W**xyzw = xyzww?"

10 · WARBLERS AND HUMMINGBIRDS

"Another bird for which I have found use is the *hummingbird* H defined by the following condition:

$$Hxyz = xyzy$$

"Show that H is derivable from B, C, and W—and hence from B, M, and T."

11 · HUMMINGBIRDS AND WARBLERS

"You can also derive a warbler from B, C, and H—in fact, you can do it from C and H, and even more simply from R and H. Can you see how?"

STARLINGS

"I have been in this forest some time now," said Craig, "and I have never seen a kestrel. Are there any kestrels here?"

"Absolutely not!" cried Bravura, in an unexpectedly fierce tone. "Kestrels are *not allowed* in this forest!!"

Craig was quite surprised at the severity of Bravura's response and was on the verge of asking him *why* kestrels were not allowed, but he decided that the question might be tactless.

"Ah, there goes a starling," Bravura said more brightly. "Tell me, are you planning to visit the Master Forest?"

"I was planning to visit Curry's Forest," replied Craig.

"And so you should!" replied Bravura. "But you shouldn't stop there; you should continue on until you reach the Master Forest. You will pass through several other interesting forests along the way—before you leave, I'll draw you a map. You will find your experience in the Master Forest to be a true education!"

"Then I'll definitely go," said Craig.

"Good!" replied Bravura. "But I should prepare you for your visit by telling you about the starling, since this bird plays a feature role in the Master Forest."

12 · STARLINGS

"A starling," said Bravura, "is a bird S satisfying the following condition:

$$Sxyz = xz(yz).\text{"}$$

"Why is that bird so important?" asked Craig.

"You will find that out when you reach the Master Forest," replied Bravura.

"Anyway," he continued, "you should know that a starling can be derived from B, T, and M—and more easily, from B, C, and W. The standard expression for S in terms of B, C, and W has seven letters, but I have discovered another having only six letters. It will be helpful to you to use the goldfinch G, which satisfies the condition $Gxyzw = xw(yz)$. The starling S is easily derivable from B, W, and G."

How is this done?

THE STARLING IN ACTION

"You have now seen that S is derivable from B, C, and W," said Bravura. "It is also possible to derive W from B, C, and S. In fact, W is derivable from just C and S, or alternatively from R and S. I will also show you that W is derivable from T and S."

13 · HUMMINGBIRDS REVISITED

"You recall that the hummingbird H is defined by the condition Hxyz = xyzy. You have seen that H is derivable from B, C, and W. We now need to find out if a hummingbird is alternatively derivable from S and C—and even more simply from S and R. Is it?"

· 14 ·

"Now write down an expression for a warbler in terms of S and R and one in terms of S and C."

"You now see," said Bravura, "that the class of birds derivable from B, C, and S is the same as the class of birds derivable from B, C, and W, since S is derivable from B, C, and W and W is derivable from C and S."

· 15 ·

"Since W is derivable from S and C, and C is derivable from B and T, then of course W is derivable from B, T, and S. However, W is derivable from just T and S. Can you show this?"

· 16 ·

"Prove that M is derivable from T and S."

"And now," said Bravura, "you see that the class of birds derivable from B, T, and W is the same as the class of birds derivable from B, T, and S, since S is derivable from B, T, and W—it is even derivable from B, C, and W, and in the other direction, W is derivable from T and S. This class of birds is also the same as the class derivable from B, M, and T, since W is derivable from B, M, and T, and in the other direction, M is derivable from W and T, as you have seen.

"More important," said Bravura, "is the fact that the class of birds derivable from B, T, M, and I is the same as the class of birds derivable from B, C, W, and I, since W is derivable from B, T, and M, and in the other direction, T is derivable from C and I—you recall that CI is a thrush. Either of these groups of four birds forms a *basis* for my forest, in the sense that every bird here is derivable from either of the foursomes. Alonzo Church preferred to take B, T, M, I as a basis; Curry preferred the basis B, C, W, I. Alternatively, we could use B, C, S, I as a basis, and for certain purposes this is technically convenient, but you will learn more about that when you reach the Master Forest."

"I am starting out tomorrow," said Craig, "and I am ever so grateful for all you have taught me. It should stand me in good stead in the journey ahead."

"It certainly should," said Bravura. "You have been a diligent student, and it has been a great pleasure to tell you some of the facts about birds I have learned. There are many more birds derivable from B, T, M, and I that I am sure would interest you. I think I will give you these derivations as exercises to take along with you to work out at your leisure. You will also encounter many other such birds in your travels ahead.

"Since you are making your journey on foot, it should take you about three days to reach Curry's Forest. This forest is named after Haskell Curry, and appropriately so, since Curry was both an eminent combinatorial logician and an avid birdwatcher. After Curry's Forest, you will come to Russell's Forest—named for Bertrand Russell. Then you will come to another forest—let's see now, I can never remember its name! Anyhow, next you will arrive at an extremely interesting forest named for Kurt Gödel. These four forests form a chain known as the Forests of Singing Birds. From Gödel's Forest it should take you two days to reach the Master Forest. I wish you the best of luck!"

Here are some of the exercises that Bravura gave to Craig. Sketches of the solutions are given at the end of the chapter.

Exercise 1 (modeled on Church's derivation of W):

a. From B and T, derive a bird G_1 satisfying the condition $G_1 xyzwv = xyv(zw)$.

b. From G_1 and M, derive a bird G_2 satisfying the condition $G_2 xyzw = xw(xw)(yz)$.

c. From B, T, and I, derive a bird I_2 such that for any bird x, $I_2 x = xII$.

d. Show that for any bird x, $I_2(Fx) = x$, where F is a finch.

e. Now show that $G_2 F(QI_2)$ is a warbler. *Note:* Q is the queer bird.

Exercise 2 (the standard starling): The standard expression for a starling in terms of B, C, and W is $B(B(BW)C)(BB)$. Show that this really is a starling.

Exercise 3: A *phoenix* is a bird Φ satisfying the condition $\Phi xyzw = x(yw)(zw)$. The bird Φ is standard in combinatory logic. Show that Φ can be derived from S and B. This is tricky! An expression of only four letters works.

Exercise 4: A psi bird is a bird Ψ satisfying the condition $\Psi xyzw = x(yz)(yw)$. The bird Ψ is also standard in combinatory logic. Show that Ψ is derivable from B, C, and W. *Hint:* Let H^* be the bird BH. The bird Ψ is easily derivable from H^* and the dovekie D_2; remember that $D_2 xyzwv = x(yz)(wv)$.

Exercise 5: It is a curious fact that Ψ is derivable from B, Φ, and—of all birds!—the kestrel K. We will divide this problem into two parts:

a. Show that from Φ and B we can get a bird Γ satisfying the condition $\Gamma\ xyzwv = y(zw)(xywv)$.

b. Show that Ψ is derivable from Γ and K.

Exercise 6: a. Show that from S and one bird already derived from B and T we can get a bird S' satisfying the condition $S'xyz = yz(xz)$.

b. Show that a warbler is derivable from S' and the identity bird I.

Exercise 7: There is a bird $\hat{Q}$ derivable from Q alone such that $C\hat{Q}W$ is a starling. Can you find it? The expression for it has six letters.

SOLUTIONS

1 • $xy(xy) = M(xy) = BMxy$, and so we take M_2 to be BM.

2 • $x(yy) = x(My) = BxMy = CBMxy$, and so CBM is a lark. Also, BxMy $= RMBxy$, and so RMB is also a lark.

We know that BBT is a robin R, and so BBTMB is a lark. Also BBTM $= B(TM)$, and so B(TM)B is a lark. This gives a fairly simple expression for L in terms of B, T, and M.

3 • $x(yy) = Bxyy = W(Bx)y = BWBxy$. Therefore BWB is a lark.

4 • $x(yy) = x(My) = QMxy$, and so QM is a lark!

5 • M_2R is a converse warbler, because $M_2Rxy = Rx(Rx)y = Rxyx = yxx$.

In terms of B, M, and R, we can take W' to be BMR. In terms of B, M, T, we can take W' to be BM(BBT).

We could also take W' to be B(BMB)T, as the reader can verify.

6 • CW' is a warbler, because $CW'xy = W'yx = xyy$. In terms of B, M, C, R, we can take W to be C(BMR). This is Bravura's expression for a warbler.

7 • We showed in Problem 22 of the last chapter that for any bird x, Cx $= B(Tx)R$. Therefore B(TW')R is a warbler. If we take BM(BBT) for W' and BBT for R, we get the expression B(T(BM(BBT)))(BBT); this expression is Bravura's. We could alternatively take B(BMB)T for W', thus getting B(T(B(BMB)T))(BBT); this is Rosser's expression for a warbler.

8 • Yes; M can even be derived from W and T, because $WTx = Txx = xx$, and so WT is a mockingbird.

We now see that the class of birds derivable from B, T, and M is the same as the class of birds derivable from B, T, and W.

9 • Take W* to be BW and W** to be B(BW).

10 • xyzy = C*xyyz = W*C*xyz. We therefore take H to be W*C*. In terms of B, C, and W, H = BW(BC).

11 • From H and R we first derive the bird W′. Well, yxx = Rxyx = HRxy. Therefore HR is a converse warbler. Hence C(HR)—or alternatively R(HR)R—is a warbler.

12 • W**G is a starling because W**Gxyz = Gxyzz = xz(yz). So we take W**G for S, which in terms of B, C, and W is the expression B(BW)(BBC).

13 • Yes, it is. SR is a hummingbird, since SRxy = Ry(xy), hence SRxyz = Ry(xy)z = xyzy.

14 • Since SR is a hummingbird, then R(SRR)R is a warbler according to Problem 11. Also C(SRR) is a warbler, and so is C(S(CC)(CC)).

15 • This is particularly simple: ST is a warbler, since STxy = Ty(xy) = xyy.

16 • We have just seen that ST is a warbler. Also, for any warbler W, the bird WT is a mockingbird, as we saw in Problem 8. Therefore STT is a mockingbird.

Solutions to the Exercises

Ex. 1: a. Take G_1 = BG.
  b. Take G_2 = G_1(BM).
  c. Take I_2 = B(TI)(TI).
We leave the last two to the reader.

Ex. 2: Left to the reader.

Ex. 3: Take $\Phi = B(BS)B$.

Ex. 4: Take $\Psi = H^*D_2$.

Ex. 5: a. Take $\Gamma = \Phi(\Phi(\Phi B))B$.

     b. $\Gamma(KK)$ is a psi bird.

Ex. 6: a. Take $S' = CS$.

     b. $S'I$ is a warbler.

Ex. 7: Take $\hat{Q} = Q(QQ(QQ))Q$.

CHAPTER 13

A GALLERY OF SAGE BIRDS

While Inspector Craig is wending his way to Curry's Forest, we will take time out to look at a medley of sage birds. But first I must tell you about *combinatorial* birds in general.

By a bird of *order 1* is meant a bird A such that for any bird x, the bird Ax can be expressed in terms of x alone. For example, the mockingbird M is of order 1, since Mx = xx and the expression xx no longer involves the letter M; it is an expression in just the letter x. Another example is the identity bird I, since Ix = x. The birds M and I are the only birds of order 1 that we have so far encountered. Of course, we could construct from the birds of the last chapter an infinite variety of birds of order 1— for example, we might wish to consider a bird A such that Ax = x(xx). The bird WL would work. Or we could construct a bird A such that Ax = (x(xx))((xxx)x)—such a bird would also be of order 1.

By a bird of *order 2* is meant a bird A such that Axy can be expressed in terms of just x and y. Examples are the thrush T, the lark L, and the warbler W; these three birds are obviously of order 2.

A bird of *order 3* is a bird A whose definition involves three variables— say, x, y, z. Thus Axyz is expressible in terms of just x, y, and z. Most of the birds we have so far encountered are of order 3—the birds B, C, R, F, and V and the queer bird Q and its relatives Q_1, Q_2, Q_3, Q_4 are all examples of birds of order 3.

We similarly define birds of order 4, 5, 6, 7, 8, and so forth. Doves are of order 4; the bald eagle Ê is of order 7.

A bird having some order or other is called a *proper combinatorial bird*—or more briefly, a *proper* bird. By a *combinatorial* bird is meant any bird expressible in terms of proper birds. Not every combinatorial bird is proper. For example, the birds T and I are both proper; hence TI is a combinatorial bird, but it is not proper, for if it were, what order could it be? It isn't of order 1, because TIx can be reduced to xI, but no further reduction is possible. TIxy can be expressed as xIy, but we haven't got rid of I, so TI is not of order 2. The best we can do with TIxyz is to express it as xIyz, but the x is still in the way, so no further reduction is possible. No matter how many variables we tack onto the right of TIxyz, we can never get rid of I, so TI is not of any order; hence it is not a proper bird. On the other hand, IT is proper, since IT = T.

SOME SAGE BIRDS

We recall that by a *sage bird* is meant a bird Θ such that for any bird x, if one calls out x to Θ, then Θ will respond by naming a bird of which x is fond—in other words, $x(\Theta x) = \Theta x$ (x is fond of Θx).

Sage birds are *not* proper birds! However, sage birds *can* be expressed in terms of proper birds; this can be done in a variety of ways that are quite fascinating. In Chapter 10 we never actually *constructed* a sage bird; we merely proved that if that forest obeyed certain conditions, then a sage bird must *exist* there. We shall now see how to *find* sage birds, given that certain proper birds are present.

· 1 ·

Derive a sage bird from a mockingbird M, a bluebird B, and a robin R. This can be done using an expression of only five letters.

· 2 ·

Find a five-letter expression for a sage bird in terms of B, C, and M.

· 3 ·

A simpler construction of a sage bird uses a mockingbird, a bluebird, and a lark. Can you find it?

· 4 ·

Derive a sage bird from a mockingbird, a bluebird, and a warbler.

· 5 ·

A tougher job is to derive a sage bird from a bluebird, a cardinal, and a warbler. Care to try it? There are several ways in which this can be done, which will become apparent in the course of this chapter.

ENTER THE QUEER BIRD

We recall that the queer bird Q satisfies the condition $Qxyz = y(xz)$. Thus Qxy composes y with x. Also $Qxyz = Byxz$. The queer bird is very useful in connection with sage birds.

· 6 ·

Show that a sage bird is derivable from a queer bird, a lark, and a warbler.

· 7 ·

Now can you see a way to solve Problem 5?

8 · QUEER BIRDS AND MOCKINGBIRDS

A particularly neat construction of a sage bird uses just the queer bird Q and the mockingbird M. Can you find it?

Discussion: By a *regular* combinator is meant a proper combinator such that, in its definition, the leftmost variable—say, x—of the left side of the equality is also the leftmost variable of the right side and occurs only once on the right side. For example, the cardinal is regular; $Cxyz = xzy$, and x is the leftmost variable of the right-hand side—xzy—and occurs only once in the expression xzy. On the other hand, the robin R is not regular; Rxyz = yzx, and x is not the leftmost variable of yzx. Also M is irregular, because x occurs twice in xx. The combinators B, C, W, L, S, I, and K are all regular; the combinators T, R, F, V, and Q are all irregular.

In each of the problems 1, 2, 3, 4, we derived a sage bird from three proper combinators; one was irregular, the mockingbird, and the other two were regular. In Problem 7 we derived a sage from three regular combinators. In Problem 8 we derived a sage from two irregular combinators, M and Q. We will now see that a sage can be derived from just two *regular* combinators—moreover, in such a fashion that each of them is derivable from B, C, and W.

CURRY'S SAGE BIRD

9 · STARLINGS AND LARKS

Show that a sage bird can be derived from a starling S and a lark L.

10 · CURRY'S SAGE BIRD

Now show that a sage bird can be derived from a bluebird, a warbler, and a starling. This can be done using an expression of only five letters.

Note: The solution of the above problem provides a second solution to Problem 5, since S can be derived from B, C, and W.

THE TURING BIRD

A bird deserving particular attention is the *Turing bird* U, defined by the following condition:

$$Uxy = y(xxy)$$

This bird was discovered by the logician Alan Turing in the year 1937, and is one of the most remarkable birds in existence! The reader will soon see why.

11 · Finding a Turing Bird

Before I tell you why I am such an admirer of the Turing bird, let's see if you *can find* one, given the birds B, M, and T, and any birds derivable from them. Can you find a Turing bird?

12 · Turing Birds and Sage Birds

The remarkable thing about the Turing bird U is that from U alone you can derive a sage bird—moreover, you can do it in as simple and direct a manner as can be imagined. Can you see how?

Some open problems: We now see that a sage bird can be derived from just *one* proper combinator—Turing's bird U. Of course, U is not regular. Can a sage be derived from just one *regular* combinator? I tend to doubt it, but I cannot prove that the answer is negative. Can a sage be derived from B and one other regular combinator? This is another question I have not been able to answer. As far as I know, these two problems are open, though I haven't checked the literature sufficiently to be sure of this.

Owls

13 · Owls

An extremely interesting bird is the *owl* O defined by the following condition:

$$Oxy = y(xy)$$

Show that an owl can be derived from B, C, and W—in fact, from just Q and W.

· 14 ·

A sage bird can be derived from O and L. Better yet, a Turing bird is derivable from O and L. How?

· 15 ·

Show that a mockingbird is derivable from O and I.

· 16 ·

Show that O is derivable from S and I.

WHY OWLS ARE SO INTERESTING

17 · A PRELIMINARY PROBLEM

Preparatory to the next problem, prove that if a bird x is fond of a bird y, then x is fond of xy.

· 18 ·

An interesting thing about owls is this: If you call out a sage bird to an owl, the owl will always respond by naming a sage bird—either the same sage bird or a different one. In other words, for any sage bird Θ, the bird $O\Theta$ is also a sage bird. Prove this.

· 19 ·

Another interesting thing about owls is that if you call out an owl to a sage bird, the sage bird will respond by naming a sage bird. In other words, for any sage bird Θ and any owl O, ΘO is a sage bird. Prove this.

· 20 ·

Equally if not more interesting is the fact that an owl is fond *only* of sage birds! In other words, for any bird A, if OA = A, then A must be a sage bird. Prove this.

· 21 ·

The last problem has as a corollary a fact that generalizes the result of Problem 19. Let us say that a bird A is *choosy* if it is fond only of sage birds. All owls are choosy, according to the last problem, but there may be other choosy birds. Now let Θ be a sage bird. Prove that it is not only the case that ΘO is a sage, as in Problem 19, but that for *any* choosy bird A, the bird ΘA must be a sage.

22 · SIMILARITY

A bird A_1 is said to be *similar* to a bird A_2 if A_1 and A_2 respond the same way to any bird x—in other words, for every bird x, $A_1 x = A_2 x$. As far as their responses to birds are concerned, *similar* birds behave identically.

We proved in Problem 18 that for any sage bird Θ, the bird $O\Theta$ is also a sage, but we didn't prove that $O\Theta$ is necessarily the same bird as Θ. However, $O\Theta$ can be proved to be *similar* to Θ. How?

Remarks: A bird forest is called *extensional* if no two distinct birds are similar—in other words, if for any birds A_1 and A_2, if A_1 is similar to A_2, then $A_1 = A_2$. Extensional forests might also be called *sparse*, since it easily follows from the extensional condition that there cannot be more than one identity bird, one mockingbird, one cardinal, one starling, and so forth.

Although the subject of extensionality is an important one, we will not be treating it in this volume. There is one fact, though, that I believe will interest you: In an *extensional* forest, an owl is fond of *all* sage birds! Do you see how to prove this?

I hope you see the ramifications of this! This fact, together with Problem 20, implies that an owl is fond of sage birds and no other birds. Thus, if you go over to an owl O and call out the name of a bird x, if O responds by calling back x, then x is a sage bird; if O calls back some bird other than x, then x is not a sage bird. So, in an extensional forest, owls seem to somehow know which birds are sage birds and which ones are not. Is this not wise of them?

· 23 ·

Prove that in an extensional forest, an owl is fond of all sage birds.

Solutions

1 • Our starting point is that any bird x is fond of the bird M(BxM), as we proved in the solution of Problem 2 of Chapter 11. And so our present problem reduces to finding a bird D such that for any bird x, Θx = M(BxM).

Well, BxM = RMBx (R is the robin), so M(BxM) = M(RMBx) = BM(RMB)x. And so we can take Θ to be BM(RMB).

Let us double-check that BM(RMB) really is a sage bird: For any bird x, BM(RMB)x = M(RMBx) = RMBx(RMBx) = BxM(RMBx) = x(M(RMBx)). Since M(RMBx) = BM(RMB)x, then x(M(RMBx)) = x(BM(RMB)x). Therefore BM(RMB)x = x(BM(RMB)x)—they are both equal to BxM(RMBx)—and so BM(RMB) is a sage.

2 • BxM = RMBx, but also BxM = CBMx, and so M(BxM) = M(CBMx) = BM(CBM)x. Since x is fond of M(BxM) and M(BxM) = BM(CBM)x, then x is fond of BM(BCM)x, and so BM(CBM) is also a sage bird.

3 • We proved in Problem 25 of Chapter 9 that x is fond of Lx(Lx), where x is any bird. Now, Lx(Lx) = M(Lx) = BMLx. Hence x is fond of BMLx, which makes BML a sage bird!

Incidentally, this provides an alternative proof for the results of the last two problems:

For *any* lark L, the bird BML is a sage. Now, CBM is a lark, according to Problem 2 of the last chapter, hence BM(CBM) is a sage, which again solves Problem 2. Also RMB is a lark, according to Problem 2 of the last chapter, and so BM(RMB) is a sage, which again solves Problem 1.

4 • Since BWB is also a lark, according to Problem 3 of the last chapter, then by the above problem, BM(BWB) is a sage bird.

5 • We will defer the solution till after the next problem.

6 • Again we use the important fact that x is fond of Lx(Lx). Now, Lx(Lx) = QL(Lx)x. Also, QL(Lx) = QL(QL)x, hence QL(Lx)x = QL(QL)xx, and so Lx(Lx) = QL(QL)xx. Furthermore, QL(QL)xx = W(QL(QL))x. This proves that Lx(Lx) = W(QL(QL))x, and since x is fond of Lx(Lx), then x is fond of W(QL(QL))x, which means that W(QL(QL)) is a sage bird.

7 • If in the above expression we take BC for Q, we get W(CBL(CBL)), which can be shortened to W(B(CBL)L). We can then take BWB for L, thus getting the expression W(B(CB(BWB))(BWB)).

Another solution will result from a later problem.

8 • Again we use the fact that x is fond of Lx(Lx), and therefore x is fond of M(Lx). Now, M(Lx) = QLMx, so x is fond of QLMx, which means that QLM is a sage bird.

We can now take QM for L, because QM is a lark, as we showed in Problem 4, Chapter 12. We thus get the expression Q(QM)M. And so Q(QM)M is a sage bird, as the reader can verify directly.

9 • It is also the case that Lx(Lx) = SLLx, and so SLL is a sage bird.

10 • We just showed that SLL is a sage. Also SLL = WSL and so WSL is a sage. Since BWB is a lark, we can take BWB for L, thus getting WS(BWB).

This is Curry's expression for a fixed point combinator.

Note: We know that B(BW)(BBC) is a starling, from Problem 12, Chapter 12, and so we can take this expression for S in WS(BWB), thus getting W(B(BW)(BBC))(BWB). This is another expression for a sage in terms of B, C, and W, and so we have another solution to Problem 5.

11 • There are many ways of going about this. Here is one. Since the forest contains B, T, and M, it also contains W, L, and Q. Now, y(xxy) = Q(xx)yy = LQxyy = W(LQx)y = BW(LQ)xy. We can therefore take U to be BW(LQ).

12 • For all x and y, Uxy = y(xxy), or what is the same thing, for all y and x, Uyx = x(yyx). We take U for y and we see that UUx = x(UUx). Therefore UU is a sage bird.

13 • y(xy) = Byxy = CBxyy = W(CBx)y = BW(CB)xy. We can therefore take O to be BW(CB).

Also, y(xy) = Qxyy = W(Qx)y = QQWxy, and so QQW is also an owl.

14 • LO is a Turing bird, since LOxy = O(xx)y = y(xxy). And so also LO(LO) is a sage bird.

15 • OIx = x(Ix) = xx, so OI is a mockingbird.

16 • SIxy = Iy(xy) = y(xy), so SI is an owl.

17 • Suppose x is fond of y. Then xy = y. Since x is fond of y and y = xy, then x is fond of xy.

18 • Suppose Θ is a sage bird; we are to show that OΘ is a sage bird.

Take any bird x. Then x is fond of Θx, since Θ is a sage. Therefore, by the last problem, x is fond of x(Θx). But x(Θx) = OΘx, and so x is fond of OΘx. Therefore OΘ is a sage bird.

19 • Suppose Θ is a sage. Then for any bird y, Θy = y(Θy), so in particular, ΘO = O(ΘO). Then for any bird x, ΘOx = O(ΘO)x = x(ΘOx). So ΘOx = x(ΘOx), or equivalently, x(ΘOx) = ΘOx, which means that x is fond of ΘOx. Therefore ΘO is a sage.

20 • Suppose OA = A. Then A = OA, hence for any bird x, Ax = OAx = x(Ax). Since Ax = x(Ax), x is fond of Ax, and so A is a sage.

21 • Suppose A is choosy and Θ is a sage. Since Θ is a sage, then A is fond of ΘA. But since A is fond only of sages, then ΘA must be a sage.

22 • Suppose Θ is a sage. Then for every bird x, Θx = x(Θx). Also OΘx = x(Θx). Therefore OΘx = Θx, since both are equal to x(Θx). Therefore OΘ is similar to Θ.

23 • Suppose the forest is extensional. Now suppose Θ is a sage. By the last problem, OΘ is similar to Θ, and since the forest is extensional, then OΘ *is* the bird Θ. Thus Θ = Θ, which means that O is fond of Θ. And so in an extensional forest, O is fond *of all* sage birds.

PART FOUR

SINGING BIRDS

CHAPTER 14

CURRY'S LIVELY BIRD FOREST

When Inspector Craig reached Curry's Forest, the first thing he did was to interview the resident bird sociologist, whose name, curiously enough, was Professor Byrd.

"In this forest," said Byrd, "certain birds sing on certain days. It has been my purpose to determine which birds sing on which days. So far, I have not been able to come to a definite solution. I have been looking for one unifying principle—one general law that would enable me to decide which birds sing on which days. Over a period of many years I have gathered an enormous amount of statistical data; I have amassed tens of thousands of facts, and aided by a high-speed computer, I have been able to amalgamate all these facts into four general laws. These four laws give *me partial* information, but I cannot see how I can determine from them exactly which birds sing on which days. I have the feeling that there should *be just one general law* that would unify these four laws— much as Newton's universal law of gravitation unified Kepler's three laws of planetary motion. But I have not been able to find it. I wonder if you could help me."

"I'll do what I can," said Craig. "What are the four laws?"

"Well, we have here a very special bird P. I do not know its species, nor does it matter. The important thing is that for any bird x and any bird y, whether the same as x or different, the following laws hold:

Law 1: If y sings on a given day, then Pxy sings on that day.

Law 2: If x doesn't sing on a given day, then Pxy sings on that day.

Law 3: If the bird x and the bird Pxy *both* sing on a given day, then y sings on that day.

Law 4: For every bird x there is a bird y such that y sings on those and only those days on which Pyx sings.

"Those are my four laws," said Byrd. "Can you unify them into one grand law?"

"I'll have to think about it," said Craig, rising. "I'll be back tomorrow and tell you if I've found anything significant."

Craig went back to the inn in which he was staying and devoted some time to the matter. At one point he burst out laughing. "What a ridiculously simple law!" thought Craig. "How could Byrd have overlooked it all these years? I think tomorrow I'll have a bit of fun with him."

Craig visited Byrd the next day.

"I've solved your problem," said Craig. "From your four laws I have been able to deduce one very general law, which in turn easily explains why the four particular laws are true."

"Wonderful!" cried Byrd. "What is this general law?"

"Rather than tell you outright, I'll give you a hint. It follows from your laws that all sparrows here sing on Tuesdays."

"Amazing!" cried Byrd. "It so happens that all sparrows here do sing on Tuesdays, but how could you have deduced this from what I have told you? I haven't said anything about sparrows or Tuesdays; what's so special about sparrows and Tuesdays?"

"Nothing special about either," replied Craig, "and this very fact should give you a hint as to what my general law is."

Byrd sank back in puzzled thought.

"Don't tell me," he said at last, "that *all* birds here sing on *all* days!"

"Exactly!" said Craig.

"Fantastic!" cried Byrd. "Why didn't this possibility ever occur to me before? But I still don't completely understand. Why does it follow from the four laws I have given you that all birds here sing on all days?"

· 1 ·

Why does it follow?

· 2 ·

Suppose we are given Byrd's first three laws, but instead of Byrd's fourth law, we are given that the forest contains a lark. Does it then follow that all the birds sing on all days? Suppose that instead of being given a lark, we are given that there is a cardinal; would it then follow that all the birds sing on all days? Suppose we are given *both* a lark *and a* cardinal; does it then follow that all the birds sing on all days?

· 3 ·

Again suppose we are given Byrd's first three laws, but we are not given the fourth. Can you find a *single* combinatorial bird whose presence would imply that all the birds sing on all days?

Discussion (to be read *after* the reader has gone through the solutions of the last three problems): The above problems are all closely related to a famous result known as *Curry's paradox*. Suppose that instead of talking about birds, we talk about propositions. And suppose that instead of talking about a bird singing or not singing on a given day, we talk about a proposition being true or false; every proposition is one or the other, but not both. For any proposition x and y, let Pxy be the proposition that either x is false or y is true, or what is the same thing, if x is true, then so is y. Then Byrd's first three laws correspond to the following three elementary laws of logic:

*Law 1:*If y is true, then Pxy is true.

Law 2: If x is false, then Pxy is true.

Law 3: If x and Pxy are both true, so is y.

Law 1 says that if y is true, then either x is false or y is true, which is obvious, because if y is true, then regardless of whether x is true or false, at least one of the propositions x and y is true—namely y. Law 2 says that if x is false, then either x is false or y is true; this is again obvious. As to

Law 3, suppose x and Pxy are both true. Since Pxy is true, then either x is false or y is true. The first alternative—x is false—doesn't hold, since x is true, so the second alternative must hold—y is true.

Now, suppose we add the following law, which corresponds to Byrd's fourth law:

Law 4: For any proposition x there is a proposition y such that the proposition y and the proposition Pyx are either both true or both false. That is, the bird y and the bird Pyx either both sing or both do not sing on a given day.

What happens if we add Law 4 to the other three laws of logic? We then get a paradox, because from the four laws 1, 2, 3, and 4 we can prove that *all* propositions are true, in exactly the same way as we proved from Byrd's four laws that all the birds sing. Obviously it is *not* the case that all propositions are true, and so the addition of Law 4 to the other three laws creates an absurdity. This is Curry's paradox.

It should be pointed out that Byrd's four laws *as applied to birds,* which Byrd did, doesn't create any paradox; it merely leads to the conclusion that all birds of the forest sing on all days, and there is no reason why this can't be. It is only when the four laws are applied—or, I should say, "misapplied"—to *propositions* in the way indicated above that a genuine paradox arises.

Suppose we now consider an arbitrary collection of entities called *objects,* and suppose we have a certain operation which applied to object x and object y yields a certain object xy. We then have what is called an *applicative system,* in which the object xy is called the result of *applying* x to y. We have been studying applicative systems for the last several chapters; our "objects" were birds and we took xy to be the response of x to y. Combinatory logic studies applicative systems with certain special properties, among which is the existence of various combinators, including C, which we have called a *cardinal,* and L, which we have called a *lark.* Now, suppose the "objects" we are studying include all propositions, both true and false, as well as other objects, the *combinators.* Suppose we have an object P such that for any *proposition x* and y, the object Pxy is the proposition that either x is false or y is true. If x and y are not both propositions, then Pxy is still a well-defined object and may or may not be a proposition, depending on the nature of x and y. Laws 1, 2, and 3, of course, hold, *provided x and y are propositions*! Also, assuming C and L are present, given any object x, there must be an object y such that y = Pyx, as we saw in the solution to Problem 2. In particular, given any *proposition* x there must be an *object* y such that y = Pyx, but this y needn't be a proposition! In fact, y *can't* be a proposition, because if it were, Pyx would also be a proposition and the same proposition as y, which would mean that Law 4 would hold and we would again run into Curry's paradox. So the way out of the paradox is to realize that given a proposition x, although the axioms of combinatory logic imply that there is some *object* y such that y = Pyx, such a y cannot be a proposition. Some of the earlier systems, which attempted to combine the logic of propositions with combinatorial logic, were careless on this point and so the systems turned out to be inconsistent. But, as Haskell Curry pointed out, the paradoxes were not the fault of combinatory logic itself, they were the result of the misapplication of combinatory logic to the logic of propositions.

SOLUTIONS

1 • Let us first observe that it follows from Byrd's first two laws that if y sings on all days on which x sings, then the bird Pxy must sing on all days. *Reason:* Suppose that y sings on all days on which x sings. Now consider any day. Either x sings on that day or it doesn't. If x doesn't, then Pxy sings on that day by Byrd's second law. Now suppose x does sing on that day. Then y also sings on that day (because of the assumption that y sings on all days on which x sings), and hence Pxy must sing on that day by Byrd's first law. This proves that regardless of whether x does or doesn't sing on that day, the bird Pxy sings on that day. Hence Pxy sings on all days.

Now we will show that given any bird x, it sings on all days. Well, by Law 4, there is a bird y that sings on those and only those days on which Pyx sings. Now, consider any day on which y sings. Pyx also sings on that day, by Law 4, and since y sings on that day, then x sings on that day, by Law 3. This proves that x sings on all days on which y sings, and hence Pyx sings on all days, by the argument of the preceding paragraph. Then, since y sings on the same days as Pyx, the bird y sings on all days. Therefore, on any day at all, the bird y and the bird Pyx both sing, hence x also sings on that day, by Law 3. This proves that x sings on all days.

2 • If we are given just L alone or just C alone, then I see no way of proving that all the birds sing on all days, but if we are given *both* C and L, then we can derive Law 4 as follows:

Since the lark L is present, then every bird is fond of at least one bird; we recall that x is fond of Lx(Lx). Now take any bird x. Then the bird CPx is fond of some bird y, which means that CPxy = y, hence y = CPxy. But also CPxy = Pyx, and so y = Pyx. Then of course y sings on the very same days as Pyx, because y *is* the bird Pyx! Thus Law 4 follows.

3 • Suppose that instead of being given the presence of both C and L, we are given that there is a bird A present satisfying the condition Axyz = x(zz)y. Then for any birds x and y, APxy = P(yy)x. Hence APx(APx) = P(APx(APx))x, and so y = Pyx, where y is the bird APx(APx).

Some Bonus Exercises

Exercise 1: Suppose we are given Byrd's first three laws but not the fourth. Prove that for any bird x, y, and z the following facts hold:
  a. Pxx sings on all days.
  b. If Py(Pyx) sings on all days, so does Pyx.
  c. If Pxy and Pyz sing on all days, so does Pxz.
  d. If Px(Pyz) sings on all days, so does P(Pxy)(Pyz).
  e. If Px(Pyz) sings on all days, so does Py(Pxz).

Exercise 2: Suppose we have a bird forest in which certain birds are called *lively.* We are not given a definition of *lively,* but we are told that there is a bird P such that the following three conditions hold:

a. For any birds x and y, if Px(Pxy) is lively, so is Pxy.

b. For any birds x and y, if x and Pxy are both lively, so is y.

c. For any bird x there is a bird y such that the birds Py(Pyx) and P(Pyx)y are both lively.

Show that all the birds of the forest are lively.

Exercise 3: The above exercise contains a somewhat stronger result than that of Problem 1 concerning Curry's Forest. Define a bird of Curry's Forest to be *lively* if it sings on all days. Then show that Byrd's four laws imply that the three conditions above all hold. It then follows from Exercise 2 that all the birds of the forest sing on all days, hence the solution of Problem 1 is a corollary of Exercise 2.

CHAPTER 15

RUSSELL'S FOREST

The next forest visited by Inspector Craig was known as Russell's Forest. Almost as soon as Craig arrived, he had an interview with a bird sociologist named McSnurd. He told McSnurd about his experiences in the last forest.

"As far as I know," said McSnurd, "we have no bird here satisfying Byrd's four laws. What we do have is a special bird a such that for any bird x, the bird ax sings on those and only those days on which xx sings. Also, for any bird x, there is a bird x′ such that for every bird y, the bird x′y sings on those and only those days on which xy does not sing. I hope this information will prove helpful."

Inspector Craig listened to this report with interest. Later that evening, sitting quietly in his room at the Bird Forest Inn, Craig reviewed the report and realized that McSnurd wasn't a very good observer, because the two facts he reported were logically incompatible.

· I ·

Why is McSnurd's report inconsistent?

Solution: It is best that we give the solution immediately. Suppose McSnurd's report were true. We consider the bird a satisfying the condition that for every bird x, ax sings on just those days on which xx sings. Then according to McSnurd's second statement, there is a bird a' such that for every bird x, a'x sings on just those days when ax doesn't sing. But the days when ax doesn't sing are just those days on which xx doesn't sing (because ax sings on the very same days as xx), and so we have a bird a' such that for every bird x, a'x sings on just those days when xx doesn't sing. Since this holds for *every* bird x, it holds when x is the bird a', and so $a'a'$ sings on those and only those days on which $a'a'$ doesn't sing, which is obviously a contradiction.

This paradox is a genuine one and is like the paradox of the barber who shaves those and only those people who don't shave themselves, or like Russell's famous paradox of the set that contains as members those and only those sets that do not contain themselves as members. Such a set would contain itself as a member if and only if it doesn't.

2 · A FOLLOW-UP

Inspector Craig was distinctly dissatisfied with Professor McSnurd, and so he asked one of the more learned inhabitants whether there were any other bird sociologists residing in Russell's Forest.

"This I do not know," was the reply, "but I do know that there is a *meta-bird-sociologist* in this forest; his name is Professor MacSnuff."

"Just what is a meta-bird-sociologist?" asked Craig in amazement.

"A meta-bird-sociologist is one who studies the sociology of bird sociologists. Professor MacSnuff is the leading authority, not on bird sociology, about which he knows nothing, but on bird sociologists. He is familiar with all the bird sociologists in the world, hence he should know which ones reside here. I suggest you contact him."

Craig expressed his thanks and then arranged an interview with Mac-Snuff.

"Yes, there is another bird sociologist here," said MacSnuff. "His name is also McSnurd. He is a brother of the McSnurd you have already interviewed."

Craig was delighted, and arranged an interview with this other McSnurd.

"Ah, yes," said McSnurd. "My brother is not always accurate; he should not have told you what he did. What he *should* have said is that there is a bird N here such that for any bird x, the bird Nx sings on those and only those days on which x does not sing. Also, this forest contains a sage bird, if that will help."

Inspector Craig thanked him and left. "Oh drat!" said Craig to himself a moment later. "This McSnurd is as bad as his brother!"

How did Craig know this?

3 · A Second Follow-up

"Isn't there any *competent* bird sociologist in this forest?" Craig asked Mac-Snuff on his second visit.

"There is only one more bird sociologist here," said MacSnuff. "His name is also McSnurd and he is the brother of the other two McSnurds."

None too hopefully, Craig arranged an appointment with the remaining McSnurd.

"Ah, yes," said the third McSnurd. "Neither of my brothers is very good at either observing or reasoning. The last McSnurd you saw was right about the sage bird; I have seen one here myself. But he was wrong about the bird N; what he *should* have told you is that there is a bird A here such that for any birds x and y, the bird Axy sings on those and only those days on which neither x nor y sings. Now you shouldn't get into any trouble."

Does the third McSnurd's story hold water?

Solutions

2 • Suppose McSnurd's report were correct. Then for every bird x, Nx $\neq$ x—that is, Nx is unequal to x—because Nx sings on just those days on which x doesn't. But since a sage bird is present, then every bird is fond of some bird, hence N is fond of some bird x, which means that Nx = x. This is a contradiction.

3 • The contradiction involved in this report is a bit more subtle and more interesting! Let us suppose the report is true. Take any bird x. Since there is a sage bird, then Ax, like every other bird, is fond of some bird y, so Axy = y. Thus y sings on those and only those days on which neither x nor y sings. If y ever sang on a given day, then neither x nor y would sing on that day, which means that y wouldn't sing on that day and we would have a contradiction. Therefore y never sings at all. Now, suppose there were some day on which x doesn't sing. Then neither x nor y sings on that day, hence Axy *does* sing on that day, and y sings on that day, contrary to the already proved fact that y never sings. Therefore x must sing on all days. And so we have proved that *every* bird x sings on all days, yet we have shown that for every bird x there is some bird y that never sings. This is obviously a contradiction.

CHAPTER 16

THE FOREST WITHOUT A NAME

Unable to find any reliable bird sociologist in Russell's Forest, Craig left it in disgust. Over the next several days he wended his weary way to the forest of this story.

For the first few days of his sojourn here, he was unaccountably sad. He could not analyze just *why* he was sad, but the fact remained that he was sad. "Could it be the disappointing results of my visit to the last forest?" thought Craig. "No," he concluded, "something else is also wrong, but I can't put my finger on just what the something is!"

Craig brightened somewhat when he heard that the bird sociologist of this forest was the eminent Professor McSnurtle. Though a cousin of the McSnurd brothers, McSnurtle was known to be thoroughly reliable. Craig had read about him back home in the *Encyclopedia of Bird Sociology,* and the one thing that was emphasized was that McSnurtle *never made mistakes*! Craig was granted an interview.

"We have a special bird e," said McSnurtle. "After years of research, I have established the following four laws concerning e.

Law 1: For any birds x and y, if exy sings on a given day, so does y.

Law 2: For any birds x and y, the bird x and the bird exy never sing on the same day.

Law 3: For any birds x and y, the bird exy sings on all days on which x doesn't sing and y does sing.

Law 4: For any bird x there is a bird y such that y sings on the same days as eyx.

"That," said McSnurtle proudly, "neatly sums up all I know about the singing habits of the birds of this forest."

Inspector Craig pondered this analysis well. At one point he could not completely suppress a slightly disdainful expression.

"What's wrong?" asked McSnurtle, who was quite a sensitive individual. "Have you found an inconsistency in my statements?"

"Oh, no," replied Craig. "I thoroughly trust your reputation for complete accuracy. Only there is one question I would like to ask you: Have you ever heard any birds in this forest sing at all?"

Professor McSnurtle wracked his brain for several minutes. "Come to think of it, I don't believe I ever have!" he finally replied.

"And I'm afraid you never will," said Craig, rising. "You could have stated your laws more succinctly still by combining them into the one simple law: *None of the birds of this forest ever sing.* I see now why I've felt so sad here!"

How did Craig realize this?

SOLUTION

This is essentially Problem 1 of Curry's Forest again. Let us say that a bird is *silent* on a given day if it doesn't sing on that day. Then McSnurtle's four laws can be equivalently stated as follows:

Law 1: If y is silent on a given day, then exy is silent on that day.

Law 2: If x is not silent on a given day, then exy is silent on that day.

Law 3: If the bird x and the bird exy are both silent on a given day, then y is silent on that day.

Law 4: For any bird x there is a bird y such that y is silent on those and only those days on which eyx is silent.

And so Byrd's four laws for P hold for e if we simply replace "sings" by "is silent." Then the same argument showing that all the birds of Curry's Forest sing on all days shows that all the birds of this forest are silent on all days.

Epilogue: Many years later, the Forest Without a Name (which actually did have a name of a paradoxical sort) came to be known as the Forest of Silence.

CHAPTER 17

GÖDEL'S FOREST

Craig's next adventure was far more delightful and also highly informative. After leaving the Forest Without a Name, he found himself in the lovely forest of this chapter. The first thing he noticed was the abundance of birds in song. They sang so beautifully—just like nightingales! The bird sociologist of this forest was a certain Professor Giuseppe Baritoni, who himself had been an excellent singer in his day.

"Now in *this* forest," explained Baritoni, "we do not regard it of much importance which birds sing on which days; the important question is which birds can sing at all! Not all birds of this forest can sing. We have plenty of nightingales, and they all sing, as you may have gathered."

"Oh, yes," said Craig. "As a matter of fact, all the birds I have heard so far have sounded to me like nightingales. Are nightingales the *only* birds here who sing, or are there others?"

"Ah, a most interesting question!" replied Baritoni. "Unfortunately we have not found the answer. The only birds I have heard sing here are nightingales, and I don't know anyone who has heard a singing bird that is not a nightingale. Still, that's not conclusive evidence that nightingales are the only singing birds of this forest; it may be that there is some bird not yet discovered that sings but is not a nightingale. It would be most interesting if there were!

"As a matter of fact, a logician from the Institute for Advanced Study in Princeton once visited this forest many years ago, and when I told him some of the singing laws of this forest, he conjectured that it *should* be decidable on the basis of these laws whether or not there was such a bird. Unfortunately, he left one day quite suddenly, and I forgot his name. I have never heard from him since."

"What *are* these laws?" asked Craig with enormous interest.

"Well," explained Baritoni, "the first interesting thing about this forest is that all the birds are married. For any bird x, by x′ I mean the mate of x. The interesting thing is that for any birds x and y, the bird x′y sings if and only if xy does not sing.

"The second interesting thing is that every bird x has a distinguished relative x* called the *associate* of x. The bird x* is such that for every bird y, the bird x*y sings if and only if x(yy) sings.

"The third thing is that there is a special bird $\mathcal{N}$ such that whenever you call the name of a nightingale to $\mathcal{N}$, $\mathcal{N}$ responds by naming a bird that sings, but if you call to $\mathcal{N}$ any bird that is not a nightingale, then $\mathcal{N}$ responds by naming a bird that doesn't sing. In other words, for any bird x, the bird $\mathcal{N}$x sings if and only if x is a nightingale."

"Very interesting," said Craig, who then took out his notebook and wrote down the following four conditions so he would not forget them.

Condition 1: All nightingales (of this forest) sing.

Condition 2: x′y sings if and only if xy doesn't sing.

Condition 3: x*y sings if and only if x(yy) sings.

Condition 4: $\mathcal{N}$x sings if and only if x is a nightingale.

Inspector Craig thanked Professor Baritoni warmly, took his leave, and spent the day ambling through this lovely forest. He retired early that evening and, curiously enough, solved the problem in his sleep close to morning. "Eureka!" he exclaimed, jumping out of bed. "I must see Baritoni immediately!" And so he dressed hurriedly, snatched a quick breakfast, and walked briskly in the direction of Baritoni's ornithological laboratory—an unusual thing for a well-bred British gentleman to do without an invitation, but Craig can surely be excused, considering his state of euphoria. He turned a sharp bend and almost walked headlong into Baritoni, who was out for his morning constitutional, humming a tune from *Aïda.*

"I have solved your problem!" exclaimed Craig exuberantly. "There *is* a bird in this forest that sings but is not a nightingale."

"Wonderful!" cried Baritoni, clapping his hands in joy. "But tell me, is there any way we can actually *find* such a bird?"

"That depends," said Craig. "To begin with, if you know how to find a bird x and how to find a bird y, do you know how to find the bird xy?"

"Not necessarily," replied Baritoni. "However, if I know how to locate x and I know the *name* of y, then I can find the bird xy: I simply go over to x and call out the name of y. Then x *names* the bird xy. Once I know the name of xy, I can find it, because I can find any bird whose name I know. It might take several hours, but it can be done."

"Good enough!" said Craig. "Next, if you know the name of a bird x, can you find out the name of its spouse x′?"

"Oh, yes; I have a complete list of all the birds I know, telling me which is mated to which."

"Also," asked Craig, "if you know the name of a bird x, are you able to find the name of its associate x*?"

"Oh, yes; I have another such list."

"Finally," asked Craig, "do you know the name of this special bird $\mathcal{N}$?"

"Of course; its name is simply the letter $\mathcal{N}$."

"Good!" said Craig. "Then I believe I can lead you to a singing bird that is not a nightingale, but from what you've said, it may take several hours."

"In that case," said Baritoni eagerly, "let's start right now. We'll stop at the lab and I'll pack us a picnic lunch."

The two spent a good part of the day on their hunt, but they were amply rewarded. Toward twilight, they found themselves in a remote, lonely, and almost unknown region of the forest, and sure enough, perched on a low branch was a bird $\mathcal{G}$ singing away ever so beautifully, and $\mathscr{G}$ was definitely *not* a nightingale. In fact, the bird belonged to a species that neither Craig nor Baritoni had ever seen or heard before.

· I ·

How did Craig know there was such a bird, and how did the two go about finding it?

Note: The bird $\mathscr{G}$ has subsequently come to be known as a *Gödelian* bird because Craig's method of finding it paralleled Gödel's method of finding a true sentence not provable in a certain axiom system. The reader interested in seeing this parallel should compare the problems of this chapter with those of chapters 14 and 15 of *The Lady or the Tiger?* The clue to the parallel is that singing birds correspond to true sentences and nightingales correspond to *provable* sentences. Thus a singing bird that is not a nightingale corresponds to a true sentence that is not provable in the axiom system under consideration.

2 · A Follow-up

The next morning Craig and Baritoni met again.

"You know," said Craig, "last night I thought of another way of finding a bird that sings but is not a nightingale. If you care to find it, we can do so, although I cannot guarantee that when we do, it might not turn out to be the same bird we found yesterday. But it may be worth a try."

Baritoni was delighted with the idea. So they spent the day in the forest and succeeded in finding a bird $\mathscr{G}_1$ that sang but was not a nightingale. As luck would have it, $\mathscr{G}_1$ turned out to be a different bird than $\mathscr{G}$, though this could not have been predicted. Can the reader explain this?

3 · The Bird Societies

Craig was enchanted with this forest and stayed for quite a while. He found out that the birds had organized several societies. A bird A is said to *represent* a set $\mathscr{S}$ of birds if for every bird x in the set $\mathscr{S}$, the bird Ax is a singing bird and for every bird x outside the set $\mathscr{S}$, the bird Ax is a nonsinging bird—in other words, for every bird x, the bird Ax sings if and only if x is a member of $\mathscr{S}$. A set of birds is called a *society* if it is represented by some bird. For example, the set of nightingales constitutes a society, because this set is represented by the bird $\mathscr{N}$.

Craig was interested in the following problem: Does the set of singing birds constitute a society? This can be answered on the basis of just Condition 2 and Condition 3 stated by Baritoni. What is the answer? Also, from just Condition 3, it can be proved that every society must either contain at least one bird that sings or lack at least one bird that doesn't sing. How is this proved, and what bearing does it have on the problem of whether the singing birds constitute a society?

SOLUTIONS

1 • They found the bird $\mathscr{G}$ in the following manner:

Baritoni already knew the name of the bird $\mathscr{N}$, hence by consulting his first list, he knew the name of $\mathscr{N}'$—the mate of $\mathscr{N}$. Then, by consulting his second list, Baritoni found the name of the bird $\mathscr{N}'^*$. To reduce clutter, let us refer to the bird $\mathscr{N}'^*$ as A. The two men next found the bird A, went up to it, and called out its own name. A responded by naming the bird AA. The two were then able to find AA. Now we prove that AA must be a bird that sings but is not a nightingale.

We let $\mathscr{G}$ be the bird AA—in other words, $\mathscr{G}$ is the bird $\mathscr{N}'^*\mathscr{N}'^*$—and we will show that $\mathscr{G}$ sings but is not a nightingale.

The bird A has the property that for any bird x, the bird Ax sings if and only if xx is not a nightingale. The reason is: $\mathscr{N}'^*$x sings if and only if $\mathscr{N}'$(xx) sings, by Condition 3, and $\mathscr{N}'$(xx) sings if and only if $\mathscr{N}$(xx) doesn't sing, which is true if and only if xx is *not a* nightingale, because $\mathscr{N}$xx does sing if and only if xx *is* a nightingale, by Condition 4. Putting these three facts together, we see that $\mathscr{N}'^*$ sings if and only if xx is not a nightingale, and since $\mathscr{N}'^*$ is the bird A, Ax sings if and only if xx is not a nightingale.

Since it is true that for *every* bird x, the bird Ax sings if and only if xx is not a nightingale, then this is true if x is the bird A, and so AA sings if and only if AA is not a nightingale. This means that either AA sings and is not a nightingale, or AA doesn't sing and is a nightingale. However, all nightingales sing, as given in Condition 1, and so the second alternative is ruled out. Therefore AA does sing, but is not a nightingale.

The credit for this clever argument is ultimately due to Kurt Gödel.

2 • Let A_1 be the bird $\mathscr{N}^{*\prime}$, rather than $\mathscr{N}^{\prime *}$. Then A_1 is not necessarily the bird A, but it also has the property that for any bird x, the bird $A_1 x$ sings if and only if $A_1 x$ is not a nightingale. We leave the verification of this to the reader.

Then it follows by the same argument that the bird $A_1 A_1$—call this bird $\mathscr{G}_1$—sings but is not a nightingale.

In summary, the bird $\mathscr{N}^{\prime *} \mathscr{N}^{\prime *}$ and the bird $\mathscr{N}^{*\prime} \mathscr{N}^{*\prime}$ are both birds that sing and neither is a nightingale.

3 • We will first prove on the basis of just Condition 3 that any society must either contain a singer or lack some non-singer.

Take any society $\mathscr{S}$. Then $\mathscr{S}$ is represented by some bird A. Now consider the bird A^*. For any bird x, the bird $A^* x$ sings if and only if A(xx) sings, according to Condition 3. Also, A(xx) sings if and only if xx is a member of $\mathscr{S}$, because A represents $\mathscr{S}$. Therefore, $A^* x$ sings if and only if xx is a member of $\mathscr{S}$. Since this is true for every bird x, then in particular, $A^* A^*$ sings if and only if $A^* A^*$ is a member of $\mathscr{S}$. And so if $A^* A^*$ does sing, then it is a member of $\mathscr{S}$, and hence $\mathscr{S}$ contains the singing bird $A^* A^*$. On the other hand, if $A^* A^*$ doesn't sing, then $A^* A^*$ is not in $\mathscr{S}$, hence $\mathscr{S}$ lacks at least one nonsinging bird—namely $A^* A^*$. This proves that every society $\mathscr{S}$ must either contain at least one singing bird or fail to contain at least one nonsinging bird.

Now, suppose the set of all singing birds formed a society; we would get the following contradiction: The set of all singing birds would be represented by some bird A. Then by Condition 2, the bird A', the mate of A, would represent the set of all birds that *don't* sing—can you see why? This means that the set of nonsinging birds forms a society, but this is impossible, since this set neither contains a singing bird nor lacks a nonsinging bird. Therefore the set of singing birds is not represented by any bird—it is not a society.

Incidentally, the solution of this problem, together with Condition 1 and Condition 4, yields an alternative proof that there is a singing bird that is not a nightingale. Since the set of singing birds doesn't constitute a society but the set of nightingales does, by Condition 4, then the two sets are not the same. But all nightingales sing, by Condition 1, hence some singing bird is not a nightingale.

PART FIVE

THE MASTER FOREST

CHAPTER 18

THE MASTER FOREST

There is only one road leading into the great Master Forest. When Craig reached the entrance, he saw an enormous sign hanging over the gate:

THE MASTER FOREST
ONLY THE ELITE ARE ALLOWED TO ENTER!

"Oh, heavens!" thought Craig. "I have no idea if they will let me in. I've never thought of myself as elite; in fact, I'm not quite sure I know what the word really means!"

At this point, an enormous sentinel blocked his way.

"Only the elite are allowed to enter!" he said in a terrible voice. "Are you one of the elite?"

"That depends on the definition of 'elite'," replied Craig. "How do you define an elite?"

"It's not how *I* define it that counts; it's how Griffin defines it."

"And who is Griffin?" asked Craig.

"Professor Charles Griffin—he is the resident bird sociologist of this forest, and he's boss around here. It's *his* definition that counts!"

"Then what is his definition?"

"Well," replied the sentinel in a softer tone, "his definition is a very liberal one. He defines an elite as anyone who wishes to enter. Do *you* wish to enter?"

"Of course!" said Craig.

"Then by definition, you're an elite and are free to enter. I'm sure Professor Griffin will be delighted to meet you. His house is a mile and a half down the road. You can't miss it; it's in the shape of an enormous bird."

"That's a relief!" thought Craig as he wended his way to the house. "I wonder why Professor Griffin instituted such a strange rule, which in fact doesn't exclude *anybody*. What sort of a chap is this Griffin, anyway?"

Well, Craig was pleasantly surprised to find Professor Griffin a most kind and hospitable fellow. He was a gentleman in his mid-sixties with long flowing white hair and a long flowing white beard. He looked somewhat like the popular image of Moses, or of God the Father.

"Welcome!" said Griffin. "I hope you will find this forest of interest."

"I have come a long way," said Craig, "and I am very curious to know what birds you have here."

"A starling and a kestrel," replied Griffin.

"That's all?" asked Craig.

"And all birds derivable from them," replied Griffin.

"Oh, that's different! Are many birds derivable from just the starling and the kestrel?"

"Very many indeed!" replied Griffin, with a subtle and rather mysterious smile. "Are you familiar with the bluebird, the dove, the blackbird, the eagle, the bunting, the dickcissel, the becard, the dovekie, the bald eagle, the warbler, the cardinal, the identity bird, the thrush, the robin, the finch, the vireo, the queer bird, the quixotic bird, the quizzical bird, the quirky bird, the quacky bird, the mockingbird, the lark, the sage bird, the Turing bird, and the owl?"

"I know them all," replied Craig, "and you mean to tell me that *all* of them are derivable from *just* the two birds S and K?"

"Indeed they are!"

Craig sank back in thought.

"Perhaps that's not too surprising," Craig said at last. "I already know that all the birds you have just mentioned are derivable from the four birds B, T, M, and I, but I didn't know that those four birds were derivable from only two combinatorial birds. Those four birds are derivable from S and K?"

"Indeed they are," replied Griffin, "and many more birds you haven't ever heard of."

"Such as?"

"Now *this* should surprise you," replied Griffin. "From just S and K you can derive *any* combinatorial bird whatsoever! And there are infinitely many combinatorial birds!"

"Fantastic!" exclaimed Craig. "Only one thing puzzles me. How can this finite forest contain *infinitely* many birds?"

"Oh, they are not necessarily all here at the same *time,*" replied Griffin. "This is an *evolving* forest, and there is an ancient legend explaining this.—Let's see, where did I put the book?"

. Professor Griffin then rummaged around in his library-study and finally brought out one of the most worn books Craig had ever seen, although it showed signs of having originally had a most beautiful, if overly ornate, binding. The book was full of remarkable ancient paintings and drawings of birds—many of which were unfamiliar. It was written in an ancient script that Craig could not identify.

"Let me translate the legend as best I can," said Griffin. "I have some knowledge of the language, but not too much. As I understand it, it goes something like this:

"In the beginning, the forest gods started the forest with just two birds—the starling S and the kestrel K. There were already humans in the forest. New birds constantly came into existence in the following manner. A human would call out the name of some already existing bird y to some existing bird x; x would then respond by calling out the name of either some existing bird or of some nonexistent bird, but the marvelous thing is that if x named a nonexisting bird, the bird would then come into being! Thus new birds were constantly generated. The forest gods were wise in starting out with the starling and the kestrel, since from these two birds all combinatorial birds can be generated.

"That is the legend," continued Griffin. "Of course, it is only a legend, but it gives one food for thought. Some ornitheological historians have likened it to the story of Adam and Eve, though which of the birds S or K is Adam and which is Eve has been a matter of bitter controversy. Male historians like to think of S as Adam, but many female historians regard this as male chauvinism. More research is needed to settle the matter definitely. Ancient Chinese historians think of S as the yang, K as the yin, and their union as the all-embracing Tao. I could say much more about the legend's literary and historical aspects, but I'd like to get back to the purely scientific side of the story."

"The legend obviously has *some* foundation in truth," Griffin went on, "since it is really a fact that all combinatorial birds are derivable from just the two birds S and K."

"How is this known?" asked Craig.

"I will reveal the secret shortly," replied Griffin, "but first I would like you to try some concrete problems."

· I ·

"Before we can get to first base," continued Griffin, "we must derive an identity bird I from S and K. Can you see how to do it?"

Inspector Craig fiddled around with this a bit, using pencil and paper, and got the solution. (The solution to this and the next three problems is incorporated into the text of "The Secret," later in this chapter.)

· 2 ·

"Good!" said Griffin. "Now that we have the identity bird I, we are free to use it in future derivations, since it is derivable in terms of S and K. Most of our future derivations will be in terms of S, K, and I.

"Next, see if you can derive a mockingbird from S, K, and I—in fact, I'll give you a hint: A mockingbird can be derived from just S and I. Can you see how?"

Craig did not have too much difficulty with this one.

· 3 ·

"Now let's take another familiar bird—the thrush," said Professor Griffin. "See if you can derive a thrush from S, K, and I."

Inspector Craig worked for quite a while on this one, but could not solve it.

"I'll give you some hints," said Griffin. "First find an expression X_1 satisfying the following two conditions:

1. X_1 is composed of just the letters S, K, I, and the variable x; the variable y should not be part of X_1.

2. The relation $X_1y = yx$ must hold, on the basis of the given conditions for S and K and also I."

Craig worked on this for a bit and found such an expression X_1.

"Now that I have it, what do I do with it?" asked Craig.

"The next step," replied Griffin, "is to find an expression X_2 having no variables at all—an expression built from just the letters S, K, and I—such that the relation $X_2x = X_1$ must hold. Then $X_2xy = X_1y$, and $X_1y = yx$ must hold, hence, the relation $X_2xy = yx$ must hold, so X_2 will be an expression for a thrush."

"Ah!" said Craig. "I begin to see light!"

· 4 ·

"Now let's try a more complex one," suggested Griffin. "Try finding a bluebird in terms of S, K, and I. There are now three variables involved—x, y, and z. First find an expression X_1 in which z doesn't occur such that the relation $X_1z = x(yz)$ must hold. Then find an expression X_2 in which neither y nor z occurs—that is, x is to be the only variable—such that the relation $X_2y = X_1$ holds. Finally, find an expression X_3 in which no variable occurs such that the relation $X_3x = X_2$ must hold. Then X_3 will be an expression for a bluebird."

THE SECRET

Before telling the reader the general method of deriving any combinatorial bird from S and K, we will first solve the four problems given by Griffin.

First we derive an identity bird from S and K (Problem 1). Well, SKK is such a bird, because for any bird x, SKKx = Kx(Kx) = x.

We might remark that for *any* bird A, the bird SKA is an identity bird (why?), so, for example, SKS is also an identity bird. But for definiteness, we will take I to be the bird SKK, which is the usual convention.

Now let's derive a mockingbird from S, K, and I (Problem 2). As Professor Griffin remarked, we can get a mockingbird from just S and I. Well, since Ix = x, it follows that SIIx = Ix(Ix) = x(Ix) = xx, and so SII is a mockingbird.

Next for the thrush. This is trickier, since there are now two variables involved—x and y. As Professor Griffin suggested, let's first find an expression X_1 whose only variable is x such that the relation X_1y = yx must hold. Well, I is an expression such that Iy = y, and Kx is an expression such that Kxy = x, hence SI(Kx) is an expression such that SI(Kx)y = yx, because SI(Kx)y = Iy(Kxy) = Iyx = yx. So SI(Kx) is an expression in which y doesn't occur and which is such that SI(Kx)y = yx. We have thus found the desired expression X_1—namely, SI(Kx).

Now we follow Professor Griffin's second suggestion and look for an expression X_2 with no variables at all such that the relation X_2x = SI(Kx) holds. Well, K(SI) is an expression such that K(SI)x = SI, and K is obviously an expression such that Kx = Kx, and so S(K(SI))K is an expression such that S(K(SI))Kx = SI(Kx). We can check this: S(K(SI))Kx = K(SI)x(Kx) = SI(Kx), since K(SI)x = SI. Therefore S(K(SI)) is a thrush. The reader can check this by computing S(K(SI))xy; he will end up with yx.

Finally, for the bluebird (Problem 4): We must first find an expression X_1 whose only variables are x and y such that the relation X_1Z = x(yz) holds. Well, since Kx is an expression such that Kxz = x and y is an expression such that yz = yz, then S(Kx)y is an expression such that

$S(Kx)yz = x(yz)$. *Check:* $S(Kx)yz = Kxz\ (yz) = x(yz)$. So X_1 can be taken to be $S(Kx)y$. It involves only the variables x and y.

Next we need an expression X_2 whose only variable is x such that the relation $X_2y = S(Kx)y$ holds. Here we have unexpected luck, since we can take X_2 to be $S(Kx)$. *Note:* The expression $S(Kx)y$ is already in the form X_2y, if y is not a variable of X_2. We are not often that lucky!

Finally, we need an expression X_3 with no variables at all such that the relation $X_3x = S(Kx)$ holds. Well, since KS is an expression such that $KSx = S$ and K is an expression such that $Kx = Kx$, then $S(KS)K$ is the expression X_3 that we seek. *Check:* $S(KS)Kx = KSx(Kx) = S(Kx)$. Therefore $S(KS)K$ must be a bluebird, as the reader can check by showing that $S(KS)Kxyz = x(yz)$.

We have by now pretty well illustrated the general method, which is this: Our expressions are built from the letters S, K, I, and variables x, y, z, w, v, and any others we might need. Let α stand for any one of the variables. For any expression X, call an expression X_1 an α-eliminate of X if the following two conditions hold:

1. The variable α does not occur in X_1.
2. The relation $X_1\alpha = X$ must hold. By this I do not mean that $X_1\alpha$ *is* necessarily the expression X, but only that the equation $X_1\alpha = X$ is derivable from the defining conditions of S and K. For example, "$KK\alpha$" and "K" are different expressions, but the relation $KK\alpha = K$ does hold, by virtue of the defining condition of the kestrel—namely that for *any* x and y, $Kxy = x$.

The fundamental problem, then, is this: Given an expression X and a variable α, how do we find an α-eliminate of X? This can always be done by a finite number of applications of the following four principles:

Principle 1: If X consists of just the variable α standing alone, then I is an α-eliminate of X. Stated otherwise, I is an α-eliminate of α. *Reason:* The variable α is obviously not part of the expression I and $I\alpha = \alpha$ does hold. Therefore I satisfies both the conditions of being an α-eliminate of α.

Principle 2: If X is an expression in which the variable α doesn't even occur, then KX is an α-eliminate of X. The reason is obvious: Since α

doesn't occur in X, then it doesn't occur in KX and the relation $KX\alpha$ = X holds.

Principle 3: If X is a composite expression $Y\alpha$ and α doesn't occur in Y, then Y itself is an α-eliminate of X. Stated otherwise, if α doesn't occur in Y, then Y is an α-eliminate of $Y\alpha$. The reasons are obvious. As an example, yz is an x-eliminate of yzx, since x doesn't occur in yz and yz is an expression E such that Ex = yzx. Also KyI is an x-eliminate of KIyx, but KIy is *not* a y-eliminate of KIyx!

Principle 4: Suppose X is a composite expression YZ, and that Y_1 is an α-eliminate of Y, and Z_1 is an α-eliminate of Z. Then the expression of SY_1Z_1 is an α-eliminate of X. *Reason:* The relations $Y_1\alpha$ = Y and $Z_1\alpha$ = Z both hold, by hypothesis, and the relation $SY_1Z_1\alpha = Y_1\alpha(Z_1\alpha)$ holds, hence the relation $SY_1Z_1\alpha$ = YZ = X holds. Also α doesn't occur in Y_1 or in Z_1—by hypothesis that Y_1 and Z_1 are respectively α-eliminates of Y and Z—hence α doesn't occur in SY_1Z_1. Hence SY_1Z_1 is an expression X_1 in which alpha doesn't occur and which has the property that the relation $X_1\alpha$ = X must hold.

We note that Principle 4 reduces the problem of finding an α-eliminate of a complex expression YZ to the problem of finding α-eliminates of each of the shorter expressions Y and Z. To find one or both of these, you might have to use Principle 4 again, and perhaps again and again, but since the expressions involved are getting shorter and shorter, the process must finally terminate.

Let us consider some examples. Suppose we wish to find an x-eliminate of the expression yx(xy). In unabbreviated notation, the expression is (yx)(xy). We see that Principle 4 is the only one that is immediately applicable, and so we must first find an x-eliminate of yx and an x-eliminate of xy. Well, by Principle 3, y is an x-eliminate of yx. As to xy, we must again use Principle 4: Since I is an x-eliminate of x and Ky is an x-eliminate of y, then by Principle 4, SI(Ky) is an x-eliminate of xy. So y is an x-eliminate of yx and SI(Ky) is an x-eliminate of xy; therefore, by Principle 4, Sy(SI(Ky)) is an x-eliminate of yx(xy). The reader can check that Sy(SI(Ky))x = yx(xy).

On the other hand, suppose we wanted to find a y-eliminate of (yx)(xy). We must first find a y-eliminate of yx and a y-eliminate of xy. As to the former, since I is a y-eliminate of y and Kx is a y-eliminate of x, then SI(Kx) is a y-eliminate of yx. As to the latter, x is a y-eliminate of xy. So SI(Kx) is a y-eliminate of yx and x is a y-eliminate of xy, hence, by Principle 4, S(SI(Kx))x is a y-eliminate of yx(xy). The reader can check that the relation S(SI(Kx))xy = yx(xy) must hold.

Now that we know how to find an α-eliminate of X, for any variable α and any expression X, we can derive from S, K, and I any combinator to do any required job. If X has only one variable—say x—and we wish to find a combinator A such that the relation Ax = X holds, we take for A any x-eliminate of X. *Example:* Suppose we want a combinator A such that the relation Ax = x(xx) holds. Well, I is an x-eliminate of x, so SII is an x-eliminate of xx. Since I is an x-eliminate of x and SII is an x-eliminate of xx, then SI(SII) is an x-eliminate of x(xx). So a combinator A that works is SI(SII), as the reader can check.

Suppose we have an expression X involving two variables—say x and y—and we seek a combinator A such that the relation Axy = X holds. We first find a y-eliminate of X—call it X_1—and then we find an x-eliminate of X_1—call it X_2, so X_2 is the combinator we seek. As an example, suppose we want a combinator A such that for any x and y, Axy = yx(xy). Well, we have already found a y-eliminate of yx(xy)— namely S(SI(Kx))x. We must then find an x-eliminate of S(SI(Kx))x. We can arrange our work as follows:

1. K(SI) is an x-eliminate of SI.

2. K is an x-eliminate of Kx.

3. Therefore S(K(SI))K is an x-eliminate of SI(Kx).

4. KS is an x-eliminate of S.

5. Hence, according to steps 4 and 3 and Principle 4, S(KS)(S(K(SI))K) is an x-eliminate of S(SI(Kx)).

6. I is an x-eliminate of x.

7. Therefore, according to steps 5 and 6 and Principle 4, S(S(KS)(S(K(SI))K))I is an x-eliminate of S(SI(Kx))x and is a combinator A doing the required job that Axy = yx(xy), as the reader can verify.

In short, if X is an expression in just two variables x and y, a combinator A that works for X—by which we mean that the relation Axy = X holds—is obtained by finding an x-eliminate of a y-eliminate of X—such an expression we call an x-y-eliminate of X. If X contains three variables x, y, and z, we find A by finding an x-eliminate of a y-eliminate of a z-eliminate of X—such an expression we call an x-y-z-eliminate of X. We have already done this for the expression x(yz), when we derived the bluebird.

Let us conclude with another example—finding a queer bird. Of course, we have already derived B and T from S and K, and in an earlier chapter we derived Q from B and T, but let us forget that we know this and see how to derive Q directly from S, K, and I.

The expression X is now y(xz). I will condense some of the steps. Thus, Ky is a z-eliminate of y; x is a z-eliminate of xz, so S(Ky)x is a z-eliminate of y(xz). Now we must find a y-eliminate of S(Ky)x. Well, S(KS)K is a y-eliminate of S(Ky)—I have condensed two steps—and Kx is a y-eliminate of x, so S(S(KS)K)(Kx) is a y-eliminate of S(Ky)x. Finally, we must find an x-eliminate of S(S(KS)K)(Kx). Well, an x-eliminate of S(S(KS)K) is K(S(S(KS)K)) and an x-eliminate of Kx is K, so S(K(S(S(KS)K))K) is an x-eliminate of S(S(KS)K)(Kx), and hence is a queer bird, as the reader can verify.

Of course, the procedure is applicable to an expression X with any number of variables. If X has four variables x, y, z, and w, we find the desired combinator by first finding the w-eliminate of X; then the z-eliminate of the result; then the y-eliminate of that result; and then the x-eliminate of *that*. Such an expression is called an x-y-z-w-eliminate of X. As an exercise, the reader should try to derive a dove D from S, K, and I. We recall that Dxyzw = xy(zw).

Some remarks are in order. First of all, the procedure we have described can be exceedingly tedious and often leads to much longer expressions than can be found by using cleverness and ingenuity. However, it is surefire, and is bound finally to result in the combinator you are seeking.

Secondly, it should be observed that the combinator you finally wind up with is not necessarily unique, because the method of finding α-eliminates can lead to several solutions, depending on the order in which you use the four principles. As an example, suppose we wish to find a z-eliminate of the expression xy. On the one hand, we can use Principle 2 and get K(xy) as a z-eliminate of xy. On the other hand, since Kx is a z-eliminate of x and Ky is a z-eliminate of y, then S(Kx)(Ky) is a z-eliminate of xy. Of course, the expression K(xy) is the simpler of the two, but K(xy) and S(Kx)(Ky) are both z-eliminates of xy, since K(xy)z = xy and also S(Kx)(Ky)z = Kxz(Kyz) = xy.

Another example: Suppose we want to find a y-eliminate of xy. On the one hand, x is a y-eliminate of xy, by Principle 3. On the other hand, since Kx is a y-eliminate of x by Principle 2 and I is a y-eliminate of y by Principle 1, then by Principle 4, S(Kx)I is also a y-eliminate of xy—a far more clumsy one than x, to be sure!

We see now that our process of finding an α-eliminate of an expression X is not deterministic; it can lead to more than one α-eliminate. It can be made completely deterministic by simply observing the following restriction: *Never use Principle 4 if any of the other three principles is applicable!*

With this restriction on the procedure, you can obtain only one α-eliminate of a given expression X. This deterministic procedure can be easily programmed on a computer, and those of you with home computers who like to work up software should find it an entertaining and profitable exercise to write a program to find combinators for any given expression.

CHAPTER 19

Aristocratic Birds

Craig spent several weeks in the Master Forest and learned many interesting things from Professor Griffin.

"You seem to have known about many birds before you ever came here," remarked Griffin in one of their daily chats. "Where did you learn about them?"

"I learned about most of them from a certain Professor Adriano Bravura. Have you heard of him?"

"Oh, heavens!" cried Griffin. "He was my teacher! I spent several years in his forest. That's where I got my start."

"Several things have puzzled me," said Craig. "Professor Bravura showed me how to derive a large number of birds from just the four birds B, T, M, and I. Are *all* combinatorial birds derivable from these four birds?"

"Definitely not," replied Griffin. "The kestrel K cannot be derived from B, T, M, and I."

"Why is that?" asked Craig.

"This can be rigorously proved," replied Griffin. "The essential idea behind the proof is this:

"The bird K has what is known as a *cancellative* effect in that Kxy = x. Look at the right side of the equation Kxy = x. What has happened to the bird y? It has mysteriously disappeared—maybe it has flown away! Anyway, we say that y has been *canceled*, and hence that K has a *cancellative* effect. Likewise the bird K_2 obeying the condition $K_2xy = y$ is cancellative; the variable x has disappeared. I could name many more cancellative birds. Now, none of the birds B, M, T, and I are cancellative, and it is impossible to derive a cancellative bird from noncancellative birds. Therefore the cancellative bird K cannot be derived from B, T, M, and I."

"That's interesting," said Craig, "and that reminds me of another thing that has puzzled me. I once asked Professor Bravura whether there were any kestrels in his forest. He seemed somewhat upset by the question and replied in a strained voice: 'No! Kestrels are not allowed in this forest!' I felt like asking him why, but the subject obviously upset him. Do you know anything about this?"

"Oh, yes," said Griffin with a laugh. "You see, Bravura is somewhat of a purist and wants only aristocratic birds in his forest."

"Now, what on earth do you mean by an aristocratic bird?" asked Craig, in amazement.

"I got the term from Bravura," replied Griffin, still amused. "By an *aristocratic* bird, he simply means a combinatorial bird that is noncancellative."

"Why the term Aristocratic'?" asked Craig.

"Well, you see, he is a bit eccentric in some of his ways. He comes from the ancient Italian nobility and has some rather old-fashioned aristocratic attitudes toward life. He regards any bird who 'cancels out' other birds as somehow lacking in nobility; he calls such birds *common* birds. The other birds he calls *aristocratic* birds.

"In a way, I can see his point," continued Griffin, "though of course I allow common birds in my forest, since they have a valuable mathematical function. And yet, if I have my choice of deriving an aristocratic bird either from S and K or from the four aristocratic birds B, T, M, and I, I tend to favor the latter. 1 am always a little uncomfortable using a common bird to derive an aristocratic one."

"Are all aristocratic birds derivable from B, T, M, and I?" asked Craig.

"Yes; there is a well-known recipe for deriving all aristocratic birds from B, T, M, and I—or more directly, from B, C, S, and I. I will show it to you sometime, if you like."

Note: The recipe is given in the appendix to this chapter.

"One thing strikes me as curious," said Craig. "From just two combinators—S and K—*all* combinators are derivable; yet we apparently need *four* aristocratic birds to derive all aristocratic birds. Why is this?"

"That's actually not true," replied Griffin. "It is true that of the four birds B, T, M, and I, no one of them is derivable from the other three. It is also true that of the four birds B, C, W, and I—which generate the same group as B, T, M, and I—none of them is derivable from the other three. It is also true that of the four birds B, C, S, and I—which again generate the same group as either of the other two foursomes—none of them is derivable from the other three. Nevertheless, there *does* exist a pair of aristocratic birds from which all aristocratic birds are derivable."

"That's interesting!" exclaimed Craig. "What are the two birds?"

"One of them is the identity bird I," replied Griffin, "and the other is a bird that may not be familiar to you—it is the *jaybird* discovered by J. Barkley Rosser in 1935. The bird J is defined by the following condition:

$$Jxyzw = xy(xwz).$$

"That is a curious bird!" said Craig. "You are right; I'm not familiar with it. Please tell me more about it."

"Very well," said Griffin. "I will first show you that J is derivable from B, T, M, and I—more directly, from B, C, W, and I. In fact, J is derivable from just B, C, and W. Then I will show you that B, T, and M are derivable from J and I. It will then follow that the class of birds derivable from J and I is the same as the class of birds derivable from B, T, M, and I."

DERIVATION OF THE JAYBIRD

1 · DERIVATION OF J

"There are several ways of deriving J from B, C, and W," said Griffin. "Perhaps the simplest uses the eagle E, the bird C*—that is, the cardinal once removed—and the bird C**—that is, the cardinal twice removed. Try to derive J from E, C*, C**, and W. Then express J in terms of B, C, and W."

GOING IN THE OTHER DIRECTION

"Now," said Griffin, "we proceed in the opposite direction. We start with the two birds J and I and set out to derive B, T, and M. We will have to rederive several familiar birds along the way, and the order will be very different from that of the original derivations from B and T. For example, we will have to derive C before B, and before *that* we must derive—of all birds!—the quixotic bird Q_1."

2 · DERIVATION OF Q_1

"You recall the quixotic bird Q_1, defined by the condition $Q_1 xyz = x(zy)$. Show that a quixotic bird is derivable from J and I."

3 · DERIVATION OF THE THRUSH

"Next, derive a thrush T from Q_1 and I."

4 · DERIVATION OF THE ROBIN

"Next, from J and T, derive the robin R."

5 · DERIVATION OF THE BLUEBIRD

"Now that we have R," said Griffin, "we can take C to be RRR, and so we have the cardinal. From C and Q_1 we can now get the bluebird B. Do you see how?

"Actually," added Griffin, "the bird C^* is easily derivable from C and Q_1, and B is derivable from C^* and Q_1. This may be the easiest path."

"Now we must derive the mockingbird," said Griffin. "This is the trickiest and most interesting part. It will be helpful first to derive a relative of J."

6 · THE BIRD J_1

"From the three birds J, B, and T, derive a bird J_1 satisfying the following condition:

$$J_1 xyzw = yx(wxz).\text{"}$$

7 · THE MOCKINGBIRD

"And now the mockingbird is derivable from C, T, and J_1. Show this.

"I'll give you a hint," added Griffin: "For any bird x, $J_1 xTTT = xx$. You can easily verify this."

Griffin continued, "Now you see that the class of birds derivable from J and I is the same as the class of birds derivable from B, T, M, and I. If we started a bird forest with just the two birds J and I, we would ultimately get the same birds as if we had started with B, T, M, and I. The kestrel K would never appear, unless it flew in from another forest, nor would a whole host of birds derivable from S and K."

Note: The theory of combinators derivable from B, T, M, and I—or equivalently from B, C, W, and I, or from just J and I—is technically known as the λ I-calculus. The theory of combinators derivable from S and K is known as the λ K-calculus. Neither theory can be said to be "better" than the other; each has applications which the other does not.

SOLUTIONS

1 • $xy(xwz)$ = $Exyxwz$ = $C^*Exxywz$ = $W(C^*E)xywz$ = $C^{**}(W(C^*E))xyzw$. And so we can take to be $C^{**}(W(C^*E))$.
In terms of B, C, and W, we have $J = B(BC)(W(BC(B(BBB))))$.

2 • We originally took Q_1 to be BCB. But in terms of J and I, we can take Q_1 to be JI, because $JIxyz = Ix(Izy) = x(Izy) = x(zy)$.

3 • We can take T to be Q_1I, because $Q_1Ixy = I(yx) = yx$.
In terms of J and I, we can take T to be JII.

4 • $JTxyz = Tx(Tzy) = Tx(yz) = yzx$. Therefore JT is a robin.
In terms of J and I, we can therefore take R to be $J(JII)$.

5 • We can take C^* to be Q_1C, because $C(Q_1C)xyzw = Q_1Cyxzw = C(xy)zw = xywz$.
Then we can take B to be C^*Q_1, because $C^*Q_1xyz = Q_1xzy = x(yz)$. Therefore C^*XQ_1 is a bluebird.
In terms of C and Q_1, $B = (Q_1C)Q_1$.

6 • $BJTxyzw = J(Tx)yzw = Txy(Txwz) = yx(wxz)$. And so we take J_1 to be BJT.

7 • First let's follow Griffin's hint. Well, $JixTTT = Tx(TxT) = Tx(Tx) = (Tx)x = Txx = xx$.
Now, $J_1xTTT = CJ_1TxTT = C(Cj_1T)TxT = C(C(CJ_1T)T)Tx$. And so we take M to be $C(C(CJ_1T)T)T$. In terms of B, C, T, and J, M = $C(C(C(BJT)T)T)T$. A bizarre expression for a mockingbird indeed! But it works.

As Curry remarked, the combinator J seems extremely artificial. It certainly does, but it yields the theoretically interesting result that the class of all combinators derivable from B, T, M, and I contains two combinators from which all the others are derivable.

APPENDIX

For the reader who is interested, here is a recipe for deriving any aristocratic bird from the four birds B, C, S, and I.

We use the notion of α-*eliminates* as defined in the last chapter. Let us call an expression X a *nice* expression if it is built up from the letters B, C, S, T, and variables. The letter K is definitely *not* allowed! What now needs to be shown is that if X is any nice expression, and if a is any variable *which actually occurs in* X, then we can find a *nice* α-*eliminate* of X. The procedure we describe for finding it will in fact lead to a *unique* nice α-eliminate of X, which we will call the *distinguished* α-eliminate of X. Our procedure is again a *recursive* one in the sense that the problem of finding the distinguished α-eliminate of a compound expression XY is sometimes reduced to the problem of first finding the distinguished α-eliminate of X and the distinguished α-eliminate of Y.

Here are the rules of the procedure.

Rule 1: The distinguished α-eliminate of α itself is I.

Rule 2: If α does not occur in X, then the distinguished α-eliminate of Xα is simply X.

Rule 3: Now consider a compound expression XY in which α occurs. Then α must occur in either X or Y and possibly in both. We assume that Y does not consist of just the variable α, because otherwise the situation reduces to Rule 2. Let X_1 be the distinguished α-eliminate of X and let Y_1 be the distinguished α-eliminate of Y. Assuming you have already found X_1 and Y_1, here is how to find the distinguished α-eliminate of XY.

a. If α occurs in both X and Y, then take SX_1Y_1 as the distinguished α-eliminate of XY.

b. If α occurs in Y but not in X, then take BXY_1 as the distinguished α-eliminate of XY.

c. If α occurs in X but not in Y, take CX_1Y as the distinguished α-eliminate of XY.

Let's consider some examples:

1. What is the distinguished z-eliminate of yz? By Rule 2 it is y.

2. What is the distinguished z-eliminate of zy? The applicable rule here is part c of Rule 3: We must first obtain the distinguished z-eliminate of z; this is I, by Rule 1. Then by part c of Rule 3, the distinguished z-eliminate of zy is CIy. We can check; CIyz = Izy = zy.

3. Find the distinguished y-eliminate of y(xy). *Solution:* I is the distinguished y-eliminate of y and x is the distinguished y-eliminate of xy, so SIx is the distinguished y-eliminate of y(xy)—according to part a of Rule 3.

4. Find the distinguished y-eliminate of z(xy). *Solution:* x is the distinguished y-eliminate of xy, so Bzx is the distinguished y-eliminate of z(xy). Checking this is obvious: Bzxy = z(xy).

Exercise: In terms of B, C, S, and I, find a combinator A satisfying the condition Axyz = xz(zy). The problem should be divided into three parts:

a. Find the distinguished z-eliminate of xz(zy).

b. Find the distinguished y-eliminate of the expression obtained in (a).

c. Find the distinguished x-eliminate of the expression found in (b). This is the desired expression A. Verify that it really works!

CHAPTER 20

CRAIG'S DISCOVERY

"I have a question for you," said Craig to Griffin the next day. "Consider the class of all birds derivable from just the three birds B, T, and I. Now—"

"Oh, that's an interesting class!" interrupted Griffin. "This class has been studied by J. B. Rosser in connection with certain logics in which duplicative birds like M, W, L, S, and J have no place. Rosser was therefore interested in the class of birds derivable from B, C, and I, but this is the same class as you have just described. Now, what do you want to know about it?"

"Can you replace the three birds B, T, and I by just *two* members of this class from which all members of the class are derivable?"

"That's an interesting question!" said Griffin. "I've never thought about it."

"Well," said Craig, "I was thinking about it last night, and I believe I've found the answer. I have discovered a bird derivable from just B and T such that both B and T are derivable from this bird together with I."

"How interesting!" cried Griffin. "What bird is that?"

"It is the goldfinch G," replied Craig. "The bird defined by the condition Gxyzw = xw(yz). Well, I have been able to derive B and T from G and I, hence the class of birds derivable from B, T, and I is the same as the class of birds derivable from G and I."

"That's neat!" said Griffin. "How do you get B and T from G and I?"

"Getting T is simple," replied Craig, "but I had a lot of trouble getting B. Here, let me show you what I have done.

· I ·

"The first thing I did was to derive the *quirky* bird—the bird Q_3 satisfying the condition $Q_3xyz = z(xy)$. This bird is easily derivable from G and I. Do you see how?"

· 2 ·

"Although this is a side issue, I might mention that the thrush T is easily derivable from Q_3 and I—hence from G and I. Can you see how?"

· 3 ·

"The important thing is to get the cardinal C," said Craig. "Can you see how to get C from G and I?"

· 4 ·

"Now that we have C," said Craig, "we have CC, which is a robin R. Then from R, G, and Q_3 we can get the queer bird Q. Do you see how? Then, of course, once we have Q and C, we have CQ, which is a bluebird B."

"Excellent!" said Griffin, after he solved these problems. "I'm delighted that after such a short time, you have begun to do original work in this field!"

SOLUTIONS

1 • GI is a quirky bird, since $GIxyz = Iz(xy) = z(xy)$. So we take Q_3 to be GI.

2 • Q_3I is a thrush, since $Q_3Ixy = y(Ix) = yx$.

3 • GGII is a cardinal, since $GGIIxyz = Gx(II)yz = GxIyz = xz(Iy) = xzy$.

4 • GRQ_3 is a queer bird, since $GRQ_3xyz = Ry(Q_3x)z = Q_3xzy = y(xz)$. So we take Q to be GRQ_3.

We might note that GR is in fact a cardinal once removed, as the reader can easily verify, and so we could take C^* to be GR. Then C^*Q_3 is a queer bird Q.

PART SIX

The Grand Question!

CHAPTER 21

THE FIXED POINT PRINCIPLE

Two days later, Craig had another session with Professor Griffin.

"Today," said Griffin, "I wish to show you an important principle known as the *fixed point principle,* which will have many applications to various topics that I plan to discuss with you later on. A special case of this principle you already know—namely, that every bird here is fond of at least one bird. Before telling you the general principle, I think it would be helpful to consider a couple of special cases. If you can solve these special cases, I'm sure you will have no trouble grasping the fixed point principle."

· 1 ·

"How do you find a bird A such that for any bird y, Ay = yA(AyA)?"

· 2 ·

"How do you find a bird A such that for any birds y and z, Ayz = (z(yA))(yAz)?"

SOLUTIONS

Inspector Craig happened to be exceptionally alert that day, and he solved the two problems in a surprisingly short time.

"I can see two different ways of going about this," said Craig. "One method uses the fact that every bird is fond of at least one bird; the other method proceeds, as it were, from scratch.

"Using the first method, here is how I solve your Problem 1. Consider the expression yx(xyx)—it is like yA(AyA) except that it has the letter x in place of the letter A. Now, by taking an x-y-eliminate of yx(xyx), we can find a bird A_1 such that for any birds x and y, $A_1xy = yx(xyx)$."

"Right, so far," said Griffin.

"Well, this bird A_1 is fond of some bird A—specifically, the bird LA_1 (LA_1), with L as a lark. Thus $A_1A = A$."

"Excellent!" said Griffin.

"Since $A_1A = A$," continued Craig, "then for any bird y, $A_1Ay = Ay$. But also, $A_1Ay = yA(AyA)$, because for *any* bird x, $A_1xy = yx(xyx)$. Since $A_1Ay = yA(AyA)$ and also $A_1Ay = Ay$, then $Ay = yA(AyA)$. This solves the problem."

"Great!" exclaimed Griffin. "But I am curious as to the second method you had in mind—the method that proceeds from scratch.' What method is that?"

"Well," replied Craig, "in the expression yx(xyx), just replace x by (xx), thus obtaining the expression y(xx)((xx)y(xx)). Then there is a bird A_2 such that for any birds x and y, $A_2xy = y(xx)((xx)y(xx))$. Then, taking A_2 for x, $A_2A_2y = y(A_2A_2)((A_2A_2)y(A_2A_2))$. And then we take for A the bird A_2A_2, and so $Ay = yA(AyA)$."

"Ah, yes!" said Griffin.

"Actually," said Craig, "I imagine the first method would, in general, yield a much shorter expression for A. The prospect of finding an x-y-eliminate of the expression y(xx)((xx)y(xx)) strikes me as pretty grim compared to finding an x-y-eliminate of the expression yx(xyx). So in practice, I think I would use the first method.

"Of course, the same method—either one, in fact—works for your second problem. To find a bird A satisfying the condition that Ayz = (z(yA))(yAz), let A_1 be an x-y-z-eliminate of the expression (z(yx))(yxz), and let A be the bird LA_1 (LA_1). Then A_1A = A, so A_1Ayz = Ayz, so Ayz = A_1Ayz = (z(yz))(yAz), and A is the desired bird.

"The same method would work for any expression with four variables instead of three. For example, take the expression x(zwy)(xxw). If we let A_1 be an x-y-z-w-eliminate of this expression and let A be the bird LA_1(LA_1), then for any birds y, z, w, Ayzw = A(zwy)(AAw). Indeed, the same method would work for any expression with *any* number of variables. Is this the principle you call the *fixed point* principle?"

"You have the idea," said Griffin. "To state the fixed point principle in its most general form, suppose we take any number of variables x,y,z . . . and write down any equation of the form Axyz . . . = (————), where (————) is any expression built from these variables and the letter A. For example, (————) might be the expression yA(wAA)(xAz). The fixed point principle is that the equation can always be solved for A—in other words, there is a bird A such that for any birds x,y,z . . . it is true that Axyz . . . = (————). In the above example, there is a bird A such that for any birds x, y, z, w it is true that Axyzw = yA(wAA)(xAz). You will see the importance of this principle when we come to the study of arithmetical birds.

"I might remark," added Griffin, "that the existence of a sage bird is only a special case of the fixed point principle—the case where (————) is the expression x(Ax). By the fixed point principle, there is then a bird A such that for every bird x, Ax = x(Ax)—such a bird A is a sage bird."

"That's interesting!" said Craig. "I hadn't seen a sage bird in that light before."

The following exercises should give the reader further insight into the uses of the fixed point principle.

Exercise 1 (Sage birds revisited): Let us look again at the problem of finding a sage bird A—only now from the point of view of the fixed point principle.

We are to find a bird A satisfying the equation Ax = x(Ax)—for all birds x. In this chapter, we have seen two different methods of solving such an equation. Try both methods and see what birds you get. Both of them have been encountered in Chapter 13.

Exercise 2 (Commuting birds revisited): Using both methods, find a bird A such that for every bird x, Ax = xA. Such a bird A *commutes* with every bird x (recall Problem 18, Chapter 11). One of the solutions will be the same as that of Problem 18, Chapter 11; the other solution will be new. What new solution do you get?

Exercise 3: In each case, find a bird A satisfying the given requirement. (Better use the first method.)
    a. Ax = Axx
    b. Ax = A(xx)
    c. Ax = AA(xx)

Exercise 4: Find a bird A such that for every bird x, Ax = AA.

Exercise 5: In each case, find a bird A satisfying the given requirement.
    a. Axy = xyA
    b. Axy = Ayx
    c. Axy = x(Ay)

Exercise 6: By the fixed point principle, there is a bird A such that for any birds x, a, and b, Axab = x(Aaab)(Abab). Using this fact, prove the following theorem (known as the *double fixed point theorem*): For any birds a and b there are birds c and d such that acd = c and bcd = d. This constitutes a new and quite simple proof of the double fixed point theorem.

SOLUTIONS

Ex. 1: Using the first method, we must first find a bird A_1 such that for all x and y, $A_1yx = x(yx)$, and any bird of which A_1 is fond will be a solution. Well, the owl O is such a bird A_1, and LO(LO)—or any other bird of which O is fond—is a sage bird. We thus get the same solution as we got in Problem 14, Chapter 13.

Using the second method, we must first find a bird A_1 such that for all x and y, $A_1yx = x(yyx)$, and then A_1A_1 will be a solution. Well, the Turing bird U is such a bird A_1, and so we see again that our old friend UU is a sage bird.

Ex. 2: Using the first method, you should get LT(LT)—or any other bird of whom the thrush T is fond—as a solution. This is the same as Problem 18, Chapter 11.

Using the second method, you should get the solution $W'W'$, where W' is the converse warbler—$W'xy = yxx$. If you get CW(CW) you are also right, since CW is a converse warbler. You can easily check that $W'W'x = x(W'W')$.

Ex. 3:
    a. LW(LW)
    b. LL(LL)
    c. L(LL)(L(LL))

Ex. 4: Some of you may have been stumped by this, since A is the only letter on the righthand side of the equation. However, either method still works; we will use the first.

We must first find a bird A_1, such that for every x and y, $A_1yx = yy$. Well, BKM is such a bird, as you can easily check, and so L(BKM)(L(BKM)) is a solution.

Ex. 5:

 a. LR(LR)
 b. LC(LC)
 c. LQ(LQ)

Ex. 6: For any birds x, a, and b, Axab = x(Aaab)(Abab). Therefore, if we take a for x, we see that Aaab = a(Aaab)(Abab). If, instead, we take b for x, we see that Abab = b(Aaab)(Abab). Therefore, if we let c = Aaab and d = Abab, we see that c = acd and d = bcd.

CHAPTER 22

A Glimpse into Infinity

Some Facts About the Kestrel

"You know," said Griffin to Craig, in another of their daily chats, "despite the fact that Professor Bravura dislikes the 'lowly' kestrel, this bird has some interesting properties."

· 1 ·

"For example," continued Professor Griffin, "suppose we have a bird forest in which there are at least two birds. You know that a kestrel cannot be fond of itself?"

"I remember that," replied Craig. He was thinking of Problem 19, Chapter 9.

"Did you know that if the forest contains at least two birds, then it is impossible for a kestrel to be fond of an identity bird?"

"I never thought about that," said Craig.

"The proof is quite easy," remarked Griffin.

What is the proof?

· 2 ·

"I hate these silly forests having only one bird," said Griffin. "In all the problems I will give you today, I am making the underlying assumption that the forest has at least two birds. "Prove that if K is a kestrel and I is an identity bird, then I $\neq$ K—in other words, no bird can be both an identity bird and a kestrel."

· 3 ·

"Another thing," said Griffin: "No starling can be fond of a kestrel. Can you see why this is so?"

· 4 ·

"It follows from this," continued Griffin, "that no starling can also be an identity bird. Can you see why?"

· 5 ·

"I see now," said Craig, "that no bird can be both a starling and an identity bird and no bird can be both a kestrel and an identity bird. Is it possible for a bird to be both a starling and a kestrel?"

"Good question!" said Griffin. "The answer is not difficult to figure out."

What is the answer? Remember, we are assuming that the forest contains at least two birds.

· 6 ·

"Here is a simple but important principle," said Griffin. "You have already agreed that no kestrel K can be fond of itself. This means that KK $\neq$ K. This fact can be generalized: For *no* bird x is it the case that Kx = K! Can you prove this?"

Note: It will be helpful to the reader to recall the cancellation law for kestrels, which we proved in Chapter 9, Problem 16—namely, that if Kx = Ky, then x = y.

· 7 ·

"Another fact," said Griffin: "We have proved that a kestrel K cannot be fond of an identity bird I. This means that KI # I. This fact can also be generalized: Prove that there cannot be any bird x such that Kx = I."

"Well," said Griffin, "I will soon tell you an extremely important fact about kestrels. But first, how about a nice cup of tea?"

"Capital idea!" said Craig.

SOME NONEGOCENTRIC BIRDS

While Craig and Griffin are taking time out for tea, let me tell you about some other nonegocentric birds. We shall assume that the forest contains the birds K and I and that K ≠ I. On this basis, we have already proved that the kestrel K cannot be egocentric; recall that by an egocentric bird is meant a bird x such that xx = x. Many other birds can also be proved nonegocentric. We shall look at a few.

· 8 ·

Prove that no bird can be both a kestrel and a thrush.

· 9 ·

Now prove that no thrush T can be egocentric.

· 10 ·

Prove that if R is a robin, then RII ≠ I. It can also be proved, by the way, that RI ≠ I and that R ≠ I. The reader might try these as exercises.

· 11 ·

Now prove that no robin R can be egocentric.

· 12 ·

Prove that no cardinal C can be egocentric.

· 13 ·

Prove that no vireo V can be egocentric. [Vxyz = zxy]

· 14 ·

Show that for any warbler W:
  a. W is not fond of I.
  b. W is not egocentric.

· 15 ·

Show that for any starling S:
  a. SI is not fond of I. It can also be shown that S is not fond of I.
  b. S is not egocentric.

· 16 ·

Prove that for any bluebird B:
  a. BKK $\neq$ KK
  b. B cannot be egocentric.

· 17 ·

Can a queer bird Q be egocentric?

The reader might have fun looking at some other familiar birds and seeing which ones can be shown to be nonegocentric. The reader might also find it a good exercise to show that of the birds B, C, W, S, R, and T, no pair can be identical—i.e., B $\neq$ C, B $\neq$ W, . . . , B $\neq$ T, C $\neq$ W, . . . , C $\neq$ T, etc.

Kestrels and Infinity

"Well," said Griffin, after they had had a delicious tea, complete with buttered crumpets, "some of the little problems I have given you about kestrels lead to a highly significant fact. Again, we consider a forest having at least two birds. Did you know that if the forest contains a kestrel K, then it must contain *infinitely* many birds?"

"That sounds most interesting!" exclaimed Craig.

"Some of my former students have given me fallacious proofs of this fact," said Griffin. "I recall that when I told this to one student, he instantly replied: 'Oh, of course! Just consider the infinite series K, KK, KKK, KKKK, . . .

"You see why this proof is fallacious?"

· 18 ·

Why is this proof fallacious?

· 19 ·

"Of course I see why the proof is fallacious," replied Craig. "However, suppose we instead take the series K, KK, K(KK), K(K(KK)), K(K(K(KK))), . . . Will that work?"

"You got it!" said Griffin.

"To tell you the truth, that was only a *guess,*" replied Craig. "I haven't really verified in my mind that all these birds are really different. For example, how do I know that K(KK) isn't really the same bird as K(K(K(K(K(KK))))))?"

"I'll give you a hint," replied Griffin. "To simplify the notation, let K_1 be the bird K; let $K_2 = KK_1$, which is KK; let $K_3 = KK_2$, which is K(KK); let $K_4 = KK_3$, which is K(K(KK)), and so forth. Thus for each number n, $K_{n+1} = KK_n$. The problem is to show that for two different numbers n and m, it cannot be that $K_n = K_m$. For example, $K_3 = K_8$ cannot hold; $K_5 = K_{17}$ cannot hold. First recall the cancellation law for kestrels: If Kx = Ky, then x = y. Then divide your proof into three steps:

Step 1: Show that for any n greater than 1, $K_1 \neq K_n$—that is, K_1 cannot be any of the birds $K_2, K_3, K_4, \ldots$

Step 2: Show that for any numbers n and m, if $K_{n+1} = K_{m+1}$, then $K_n = K_m$. For example, if K_4 were equal to K_7, then K_3 would have to be equal to K_6.

Step 3: Using Step 1 and Step 2, show that for no two *distinct* numbers m and n can it be the case that $K_n = K_m$, and therefore there really are infinitely many birds in the sequence $K_1, K_2, K_3, \ldots$"

With these hints, Craig solved the problem. What is the solution?

SOLUTIONS

1 • Suppose a kestrel K is fond of an identity bird I. Then KI = I. Therefore, for any bird x, KIx = Ix, and since Ix = x, then KIx = x. Also KIx = I, since K is a kestrel. This means that KIx is equal to both x and I, hence x = I. Therefore, if K is fond of I, then *every* bird x is equal to I and hence I is the only bird in the forest. But we are given that there are at least two birds in the forest; hence K cannot be fond of I.

2 • This follows from the last problem. Suppose K = I. Then KI = II, hence KI = I. This means that K is fond of I, which, according to the last problem, cannot happen.

3 • Suppose SK = K. Then for any birds x and y, SKxy = Kxy. Hence SKxy = x, since Kxy = x. Also, SKxy = Ky(xy) = y. Therefore SKxy is equal to both x and y, hence x and y are equal. So, if SK = K, then any birds x and y are equal, which means that there is only one bird in the forest.

4 • Suppose S = I. Then SK = IK = K, hence SK = K. But SK $\neq$ K, as we showed in the last problem; therefore S $\neq$ I.

5 • Suppose S = K. Then SIKI = KIKI. Now, SIKI = II(KI) = I(KI) = KI, whereas KIKI = II = I. Therefore, if S = K, then KI = I. But KI $\neq$ I, by Problem 1; hence S $\neq$ K.

6 • Suppose there were a bird x such that Kx = K. Then for every bird y, it would follow that Kxy = Ky, and hence that x = Ky. Then for any birds y_1 and y_2, it would follow that Ky_1 = Ky_2, because x is equal to each of them. Then by the cancellation law—Problem 16, Chapter 9—it would follow that y_1 = y_2. And so the assumption that there is a bird x such that Kx = K leads to the conclusion that for any birds y_1 and y_2, the bird y_1 is equal to y_2—in other words, that there is only one bird in the forest!

7 • This is easier. Suppose there is a bird x such that Kx = I. Then KxI = II, hence x = I, since KxI = x and II = I. Then, since Kx = I and x = I, it follows that KI = I. But KI $\neq$ I, according to Problem 1. Therefore, there is no bird x such that Kx = I.

8 • Suppose T = K. Then TIK = KIK, hence KI = KIK = I, but KI $\neq$ I, according to Problem 7.

9 • For this and the next several problems, I will make the solutions more condensed. By now, the reader should have enough experience to fill in any missing steps. I will illustrate what I mean by "missing steps" in the solution to this problem.

Suppose TT = T. Then TTKI = TKI, hence KTI = IK. *Missing steps:* "because TTKI = KTI and TKI = IK." Therefore T = K. *Missing steps:* "because KTI = T and IK = K." But T $\neq$ K according to Problem 8. Therefore it cannot be that TT = T.

10 • Suppose RII = I. Then RHK = IK, hence IKI = K by simplifying both sides of the equation, hence KI = K, contrary to Problem 7. In Problem 7 we proved that there is *no* bird x such that Kx = K, so in particular, KI $\neq$ K.

11 • Suppose RR = R. Then RRII = RII. Now, RRII = IIR = R, so R = RII, hence RII = RIIII = IIII = I. We then have RII = I, contrary to the last problem.

12 • Suppose CC = C. Then CCIKI = CIKI = IIK = K. Also, CCIKI = CKII = KII = I. We then have I = K, contrary to Problem 2.

13 • Suppose VV = V. Then VVIII = VIII = III = I. Also VVIII = IVII = VII, and so VII = I. Then VIIK = IK = K. Also VIIK = KII = I, and so we have K = I, contrary to Problem 2.

14 • a. Suppose WI = I. Then WIK = IK = K. Then IKK = K, hence KK = K, which we know is not so; no kestrel is egocentric.

  b. Suppose WW = W. Then WWI = WI. Now, WWI = WII = III = I. Hence we would have WI = I, contrary to part a of the problem.

15 • a. Suppose SI were fond of I. Then SII = I. Then SIIK = IK, hence IK(IK) = IK, so KK = K. But KK ≠ K, so SII ≠ I.

  b. Suppose SS = S. Then SSIII = SIII = II(II) = I. Also SSIII = SI(SI)I = II(SII) = SII. Hence we have SII = I, contrary to part a of the problem.

16 • a. Suppose BKK = KK. Then BKKI = KKI, hence K(KI) = K. This is again contrary to Problem 6, which states that there is *no* bird x such that Kx = K.

  b. Suppose BB = B. Then BBIK = BIK, hence B(IK) = BIK. Therefore BK = BIK. Therefore BKK = BIKK = I(KK) = KK and we have BKK = KK, contrary to part a of the problem.

17 • Suppose QQ = Q. Then QQIKI = QIKI = K(II) = KI. Also, QQIKI = I(QK)I = QKI. Hence QKI = KI. Then QKII = KII, so I(KI) = I, hence KI = I, contrary to Problem 1. Therefore, Q, queer as it may be, is definitely *not* egocentric.

18 • The fallacy is that all the infinitely many of the expressions of the series name only two different birds—namely K and KK. Clearly KKK = K, hence KKKKK = KKK = K, and indeed all the expressions with an odd number of K's boil down to K; all those with an even number of K's boil down to KK.

19 • The series Craig named really works!

Step 1: We proved in Problem 6 that for every bird x, $K \neq Kx$. Hence K cannot be any of the birds KK_1, KK_2, KK_3, . . . Thus K_1 is not any of the birds K_2, K_3, K_4 . . .

Step 2: Suppose, for example, that $K_3 = K_{10}$. Then $KK_2 = KK_9$, hence by the cancellation law for kestrels, $K_2 = K_9$.

Of course the proof works for any numbers n and m: If $K_{n+1} = K_{m+1}$ then $KK_n = KK_m$, and so $K_n = K_m$.

Step 3: Suppose, for example, that $K_4 = K_{10}$. Then by successively applying Step 2, we would have $K_3 = K_9$, $K_2 = K_8$, $K_1 = K_7$, violating Step 1.

Obviously the proof works for any two distinct numbers.

CHAPTER 23

LOGICAL BIRDS

"I am very proud of this forest," said Professor Griffin one day. "Some of the birds here can do very clever things. For example, did you know that some of them can do propositional logic?"

"I am not sure I understand what you mean by that," replied Griffin.

"Let me first explain some of the basics of propositional logic," said Griffin. "To begin with, I am using *Aristotelian* logic, according to which every proposition p is either true or false but not both. We use the symbol t to stand for *truth* and f to stand for *falsehood*. And so the value of any proposition p is either t or f—t if p is true and f if p is false. Now, logicians have a way of constructing more complex propositions out of simpler ones. For example, given any proposition p, there is the proposition *not p*—symbolized ~p—which is false when p is true and true when p is false. This is simply schematized: ~t = f; ~f = t. It is usually displayed as the following table, called the *truth table* for negation:

	p	~p
Negation	t	f
	f	t

"Next, given any propositions p and q, we can form their *conjunction*—the proposition that p and q are both true. This proposition is symbolized p & q. It is true when p and q are both true, and false otherwise.

198

In other words, t & t = t; t & f = f; f & t = f; and f & f = f. These four conditions are tabulated by the following table—the so-called truth table for conjunction:

Conjunction

p	q	p & q
t	t	t
t	f	f
f	t	f
f	f	f

"Also, given propositions p and q, we can form the proposition p ∨ q, which is read 'p or q, or maybe both' and is called the *disjunction* of p and q. This proposition is true if *at least* one of the propositions p and q is true; otherwise it is false. The disjunction operation has the following truth table:

Disjunction

p	q	p ∨ q
t	t	t
t	f	t
f	t	t
f	f	f

"As you see, the proposition p ∨ q is false only in the last of the four possible cases—the case when p and q both have the value f.

"Next, from propositions p and q we can form the so-called *conditional* proposition p → q, which is read 'if p, then q,' or 'p implies q.' The proposition p → q is taken to be *true* if either p is false or p and q are both true. The only case when p → q is false is when p is true or q is false. Here is the truth table for p → q:

Conditional

p	q	p → q
t	t	t
t	f	f
f	t	t
f	f	t

199

"Since p → q is true when and only when p is false or p and q are both true, it can also be written: (~p) ∨ (p & q). It can be written even more simply as (~p) ∨ q, or as ~(p & ~q).

"Finally, given any propositions p and q, there is the proposition p ↔ q, which is read 'p if and only if q,' which asserts that p implies q and q implies p. This proposition is true just in the case that p and q both have the value t or both have the value f.

Equivalence

p	q	p ↔ q
t	t	t
t	f	f
f	t	f
f	f	t

"These five symbols— ~ (not), & (and), ∨ (or), → (if—then), ↔ (if and only if)—are called *logical connectives*. Using them, one can form from simple propositions propositions of any complexity. For example, we can form the proposition p & (q ∨ r), which is true if and only if p is true and also at least one of q and r is true. Or we could form the very different proposition (p & q) ∨ r, which is true just in case either p and q are both true, or r is true. One can easily compute their truth values, given the truth values of p, q, and r, by combining the tables for & and ∨. Of course, since there are now three variables involved—p, q, and r—we now have eight possibilities instead of four. Here is the truth table for (p & q) ∨ r.

p	q	r	(p & q)	(p & q) ∨ r
t	t	t	t	t
t	t	f	t	t
t	f	t	f	t
t	f	f	f	f
f	t	t	f	t
f	t	f	f	f
f	f	t	f	t
f	f	f	f	f

"On the other hand, here is the truth table for p & (q ∨ r).

p	q	r	(q ∨ r)	p & (q ∨ r)
t	t	t	t	t
t	t	f	t	t
t	f	t	t	t
t	f	f	f	f
f	t	t	t	f
f	t	f	t	f
f	f	t	t	f
f	f	f	f	f

"You see, the two propositions have different truth tables," said Griffin.

"I understand all this," said Craig, "but how does it relate to the birds?"

"I am coming to that," replied Griffin. "To begin with, I have chosen for t and f two *particular* birds. The first, t, represents *truth*, or it can be thought of as being the representative of all true propositions. The second bird, f, of course, represents *falsehood*, or is the representative of all false propositions. I call t the *bird of truth*, or the *truth bird*, or more briefly, just *truth*. I call f the *falsehood bird*, or the *bird of falsehood*, or more briefly, just *falsehood*."

"What birds are they?" asked Craig.

"For t, I take the kestrel K; for f, I take the bird KI. And so, when we are discussing propositional logic, I use t synonymously with K and f synonymously with KI."

"Why this particular choice?" asked Craig. "It seems quite arbitrary!"

"Oh, there are many other choices that would work," replied Griffin, "but this particular one is technically convenient. I have adopted this idea from the logician Henk Barendregt. I will tell you the technical advantage in a moment.

"The birds t and f are collectively called *propositional birds*. Thus, there are only two propositional birds—t and f. From now on, I shall use the letters p, q, r, and s as standing for arbitrary *propositional birds,* rather than propositions. I call p *true* if p is t and *false* if p = f. Thus t is called *true* and f is called *false*.

"Now, the advantage of Barendregt's scheme is this:

"For any birds x and y, whether propositional birds or not, txy = x, since Kxy = x, and fxy = y, since fxy = KIxy = Iy = y. And so for any *propositional* bird p, pxy is x if p is true, and pxy is y if p is false. In particular, if p. q, and r are all propositional birds, then pqr = (p & q) ∨ (~p & r)—or what is the same thing, pqr = (p → q) & (~p → r). This can be read 'if p then q; otherwise r.' "

"You still haven't told me what you mean when you say that some of the birds here can *do* propositional logic," said Craig. "Just what do you mean by this?"

"I was just coming to that!" replied Griffin. "What I mean is that for any simple or compound truth table, there is a bird here that can compute that table."

· 1 ·

"For example, there is a bird N—called the *negation* bird—that can compute the truth table for negation. That is, if you call t to N, N will respond by naming f; if you call f to N, N will respond by naming t. Thus Nt = f and Nf = t. In other words, for any propositional bird p, Np is the bird ~p. The first problem I want you to try is to find a negation bird N."

· 2 ·

"Then we have a *conjunction bird* c such that for any propositional birds p and q, cpq = p & q. In other words, ctt = t; ctf = f; cft = f; and cff = f. Can you find a conjunction bird c?"

· 3 ·

"Now find a *disjunction* bird d—a bird such that for any propositional birds p and q, dpq = p ∨ q.In other words, dtt = t; dtf = t; dft = t; but dff = f. Can you find such a bird d?"

· 4 ·

"Then there is the *if-then* bird—a bird i such that itt = t; itf = f; ift = t; and iff = t. In other words, ipq = p → q. Can you find an if-then bird i?"

· 5 ·

"Now find the *if-and-only-if bird* e—also called an *equivalence* bird—such that for any propositional birds p, q, epq = (p ↔ q). In other words, ett = t; etf = f; eft = f; and eff = t."

SOLUTIONS

1 • Since the Master Forest is combinatorially complete, we can find a bird N such that for all x, Nx = xft. Specifically, we can take N to be Vft, where V is the vireo. Then Vftx = xft. So Nt = tft = f; Nf = fft = t. Thus N is a negation bird.

2 • Consider c such that for any x and y, cxy = xyf. *Note:* We can take c to be Rf, where R is the robin. Then Rfxy = xyf.
    1. ctt = ttf = t
    2. ctf = tff = f
    3. cft = ftf = f
    4. cff = fff = f
    Thus c is a conjunction bird.

3 • Take d such that for all x and y, dxy = xty. We can specifically take d to be Tt, where T is the thrush. Then Ttxy = xty. The reader can verify that d is a disjunction bird by working out the four cases.

4 • Take i such that ixy = xyt. We can take i to be Rt, where R is the robin. The reader can verify that this bird i works.

5 • Take e to be such that for all x and y, exy = xy(Ny). We can take e to be CSN, where C is the cardinal, S is the starling, and N is the negation bird. The reader can easily verify that cpq = p ↔ q.

CHAPTER 24

BIRDS THAT CAN DO ARITHMETIC

In this episode and the next, Craig found out the true wonders of Griffin's forest.

Shortly before his departure, Craig visited Griffin in his study one late-summer day. The weather was beautiful, and all the windows of the study were open. Craig was quite surprised to see several birds perched on the windowsills engaged in lively conversation with Professor Griffin— all in bird language, of course. As the birds already there left, others would come.

"Ah yes!" said Griffin, after the last bird had departed. "I have been testing some of my arithmetical birds. Did you know that some of the birds here can do arithmetic?"

"Will you please explain that?" asked Craig.

"Well, I'd better start at the beginning," replied Griffin. "We will work with the natural number series 0, 1, 2, 3, 4 . . . When I use the word 'number' I will always mean either 0 or one of the positive whole numbers. These numbers are called *natural* numbers. By the *successor* n^+ of a number n, I mean n + 1. Thus $0^+ = 1$; $1^+ = 2$; $2^+ = 3$, and so forth.

"Now each number n is represented by some bird; I use the notation $\bar{n}$ to mean the bird that represents n. Thus n is a *number,* $\bar{n}$ is a bird—the bird that represents the number n. In the scheme I am about to show you for representing numbers by birds, the vireo V plays a major role: We will let σ be the bird Vf—which is V(KI)—and we will call σ the *successor* bird. For $\bar{0}$, we take the identity bird I. We take $\bar{1}$ to be the bird $\sigma\bar{0}$; $\bar{2}$ to be $\sigma\bar{1}$; $\bar{3}$ to be $\sigma\bar{2}$, and so forth. Hence $\bar{0} = I$; $\bar{1} = \sigma\bar{0}$; $\bar{2} = \sigma(\sigma\bar{0})$; $\bar{3} = \sigma(\sigma(\sigma\bar{0}))$, and so forth. Thus $\bar{0} = I$; $\bar{1} = Vfl$; $\bar{2} = Vf(Vfl)$; $\bar{3} = Vf(Vf(Vfl))$, and so on."

"Again, this choice strikes me as arbitrary," said Craig. "What's so special about the bird Vf?"

"You will see that shortly," replied Griffin. "Actually there are many other possible choices. The first numerical scheme was proposed by Alonzo Church. The scheme I am using has several technical advantages over Church's; the combinatorial logician Henk Barendregt is responsible for it. Anyway, I want to start explaining to you how birds here do arithmetic. First for some preliminaries:

"The birds $\bar{0}$, $\bar{1}$, $\bar{2}$, $\bar{3}$, and so on I call *numerical* birds—these are identified with the respective numbers 0, 1, 2, 3, . . . Now, if I call out a numerical bird $\bar{n}$ to a bird A, A doesn't necessarily respond by calling back a *numerical* bird; it might call back a nonnumerical bird. Well, a bird A is said to be an *arithmetical bird of type 1* if for every numerical bird $\bar{n}$, the bird A$\bar{n}$ is also a numerical bird. Loosely speaking, this means that A operating on a number gives you a number. A bird A is called an *arithmetical bird of type 2* if for any numbers n and m, the bird A$\bar{n}\bar{m}$ is a numerical bird. Equivalently, A is a numerical bird of type 2 if for every number n, the bird A$\bar{n}$ is an arithmetical bird of type 1. Similarly, we define arithmetical birds of types 3, 4, 5, and so on. Thus, for example, if A is an arithmetical bird of type 4, then for any numbers a, b, c, and d, the bird A$\bar{a}\bar{b}\bar{c}\bar{d}$ is a numerical bird.

"Now come some interesting things. There is a bird here called the *addition* bird, symbolized by $\oplus$, such that for any numbers m and n, $\oplus\bar{m}\bar{n}$ is the sum of m and n—or rather, the numerical bird representing that sum. That is, $\oplus\bar{m}\bar{n} = \overline{m+n}$. Thus, for example, $\oplus\bar{2}\,\bar{3} = \bar{5}$; $\oplus\bar{3}\,\bar{9} = \overline{12}$.

"Then we have a bird $\otimes$ called a *multiplication* bird such that for any numbers n and m, $\otimes\bar{n}\bar{m}$ is the bird n $\cdot$ m. So, for example, $\otimes\bar{2}\,\bar{5} = \overline{10}$; $\otimes\bar{3}\,\bar{7} = \overline{21}$.

"We also have an *exponentiating* bird Ⓔ such that for any numbers n and m, Ⓔn̄m̄ = k̄, where k is the number n^m—the result of multiplying n by itself m times. So, for example, Ⓔ5̄ 2̄ = 2̄5̄; Ⓔ2̄ 5̄ = 3̄2̄; Ⓔ2̄ 3̄ = 8̄; Ⓔ3̄ 2̄ = 9̄.

"Having these birds," continued Griffin," we can easily combine them to form any arithmetical combination we want. For example, we can find a bird A such that for any numbers a, b, and c, Aāb̄c̄ = d̄, where d, say, is $(3a^2b + 4ca)^5 + 7$.

"In fact," continued Griffin, in growing excitement, "given *any* numerical operation that can be performed by one of these modern electronic computers, there is a bird here that can perform the same operation! For any computer, there is a bird here that can match it!

"Do you realize what this means?" asked Griffin, waxing more excited still. "It means that the birds here could totally take over the job of the computers. Maybe one day the computers of the world will one by one be replaced by birds until there are no computers left—only birds! Wouldn't that be a beautiful world?"

Craig thought this idea somewhat visionary, but intriguing, nevertheless.

"All this sounds most interesting," said Craig, "but I am in the dark as to how you find even the basic arithmetic birds that add, multiply, and exponentiate. What birds are they?"

"I am coming to that," replied Griffin, "but first for some preliminaries."

· I ·

"To begin with," said Griffin, "we should be sure that the birds 0̄, 1̄, 2̄, 3̄, ... are all distinct—that is, for any numbers n and m, if n ≠ m, meaning n is unequal to m, then the bird n̄ is distinct from the bird m̄. Can you see how to prove this?"

2 · THE PREDECESSOR BIRD P

"For any *positive* number n," said Griffin, "by its *predecessor* n^- is meant the next lower number. That is, for any positive n, n^- is the number n - 1. Of course, for any number n, the number n^+ is positive and the predecessor of n^+ is n.

"What we now need," said Griffin, "is a bird that calculates predecessors. That is, we want a bird P such that for any number n, $\overline{Pn^+} = \bar{n}$. Can you see how to find such a bird P?"

· 3 ·

"We recall the propositional birds t and f. We now need a bird Z called a *Zero-tester* such that if $\bar{0}$ is called out to Z, you will get the response t—meaning, 'True, the number you called is 0'—whereas if you call out any number other than 0, you will get the response f—meaning, 'False, the number is not 0.' That is, we want a bird Z such that $Z\bar{0} = t$, but for any *positive* number n, $Z\bar{n} = f$. Can you find such a bird Z?"

· 4 ·

"Let me ask you a question," said Griffin. "Do you have any reason to believe that there is a bird A such that for any number n and any birds x and y, if n = 0, then $A\bar{n}xy = x$, but if n is positive, $A\bar{n}xy = y$? That is, is there a bird A such that $A\bar{0}xy = x$; $A\bar{1}xy = y$; $A\bar{2}xy = y$; $A\bar{3}xy = y$; and so forth?"

"Oh, of course!" replied Craig, after a moment's thought.

How did Craig realize this?

"And now," said Griffin, "we come to some of the more interesting birds. Before we consider the problem of finding an addition bird, let us consider a slightly simpler problem. Let us take any particular number—say 5. How can we find a bird A that adds 5 to any number that you call to it? That is, we want a bird A such that $A\bar{0} = \bar{5}$; $A\bar{1} = \bar{6}$; $A\bar{2} = \bar{7}$—and for any number n, $A\bar{n} = \overline{n + 5}$."

Craig thought about this, but could not find a solution.

· "The idea is based on a principle known as the *recursion* principle," said Griffin. "Suppose A is a bird such that the following two conditions hold:

1. $A\overline{0} = \overline{5}$

2. For every number n, $A\overline{n^+} = \sigma(A\overline{n})$.

"Do you see that such a bird A would do the required job?"

"Let us see now," said Craig. "It is given that $A\overline{0} = 5$. What about $A\overline{1}$? Well, by the second condition, $A\overline{1} = \sigma(A\overline{0}) = \sigma(\overline{5})$, since $A\overline{0} = \overline{5}$, and $\sigma\overline{5} = \overline{6}$. Therefore $A\overline{1} = \overline{6}$. Now that we know that $A\overline{1} = \overline{6}$, it follows that $A\overline{2} = \overline{7}$, because $A\overline{2} = \sigma(A\overline{1}) = \sigma\overline{6} = \overline{7}$. Yes, of course I see why it is that for every number n, $A\overline{n} = \overline{n+5}$. We successively prove $A\overline{0} = \overline{5}$, $A\overline{1} = \overline{6}$, $A\overline{2} = \overline{7}$, $A\overline{3} = \overline{8}$, and so forth!"

"Good!" said Griffin. "You have grasped the recursion principle."

"I am still in the dark, though, about how one finds a bird A satisfying those conditions," said Craig. "How does one?"

"Ah, that's the clever part," said Griffin with a smile. "It is based on the fixed point principle, which I have already explained to you."

"Really!" said Craig in amazement. "I can't see any connection between the two!"

"I will now explain," said Griffin. "First of all, do you see that Condition 2 can be alternately described as follows?

$2'$. For every number n greater than 0, $A\overline{n} = \sigma(A(P\overline{n}))$."

"Yes," said Craig, "because for any number n greater than zero, n = m^+, where m is the predecessor of n. Therefore Condition $2'$ says that $A\overline{m^+} = \sigma(A(P\overline{m^+}))$, but since $P\overline{m^+} = \overline{m}$, then Condition $2'$ simply says that $A\overline{m^+} = \sigma(A(P\overline{m^+}))$, or what is the same thing, $A\overline{n} = \sigma(A(P\overline{n}))$. But, of course, this holds only when n is positive."

"Good!" said Griffin. "And so you see that what we want is a bird A such that $A\overline{n} = \overline{5}$ if n = 0, and $A\overline{n} = \sigma(A(P\overline{n}))$ if $n \neq 0$."

"I see that," said Craig.

"Well, we use the zero-tester Z," said Griffin. "The bird $Z\overline{n}\overline{5}(\sigma(A(P\overline{n})))$ is $\overline{5}$ if n = 0, and is $\sigma(A(P\overline{n}))$ if $n \neq 0$, and so we want a bird A such that for every number n, $A\overline{n} = Z\overline{n}\overline{5}(\sigma(A(P\overline{n})))$. Well, by the fixed point principle, there *is* such a bird A—in fact, there is a bird A such that for *any* bird x, whether a numerical bird or not, Ax = $Zx\bar{5}(\sigma(A(Px)))$. That solves the problem!

"In case you have forgotten," added Griffin, "we can obtain the bird A by first taking a bird A_1 such that for any birds x and y, $A_1yx = Zx\bar{5}(\sigma(y(Px)))$, and then you can take for A any bird of which A_1 is fond—for example, we can take $LA_1(LA_1)$ for A."

"That is indeed clever!" said Craig in genuine admiration. "Who thought of it?"

"The idea of using the fixed point principle to solve problems like this is attributable to Alan Turing—the same logician who discovered the Turing bird. Turing has done some extremely clever things!"

· 5 ·

"Of course," said Griffin, "the number 5 has no special significance; I could have taken, say, 7, and asked for a bird A such that for all n, $A\bar{n} = \overline{n+7}$. However, we want something better than that. We want an arithmetic bird $\oplus$ of type 2 such that for *any* two numbers n and m, $\oplus\bar{m}\bar{n} = \overline{m+n}$. Only a slight modification of what I have shown you is necessary. Can you see how to find such a bird $\oplus$?"

· 6 ·

"Next, can you see how to find a bird $\otimes$ such that for any numbers n and m, $\otimes\bar{n}\bar{m} = \overline{n \cdot m}$? Of course, you are free to use the bird $\oplus$ that you have just found."

· 7 ·

"Now that we have the birds $\oplus$ and $\otimes$, can you find an exponentiating bird Ⓔ such that for any numbers n and m, $Ⓔ\bar{n}\bar{m} = \bar{k}$, where $k = n^m$?"

PREPARATION FOR THE FINALE

"I understand you must leave this forest in a couple of days. Is that correct?" asked Griffin.

"Alas, yes!" replied Craig. "I have been called back home on a strange case involving a bat and a Norwegian maid."

"That *does* sound strange!" remarked Griffin. "At any rate, tomorrow I would like to tell you one of the most interesting facts of all about this forest. This fact is related to Gödel's famous incompleteness theorem, as well as to some results discovered by Church and Turing. But today, I must give you the necessary background. I must tell you more about arithmetical birds as well as something about property birds and relational birds."

"What are *they*?" asked Craig.

· 8 ·

"Well, by a *property* bird is meant a bird A such that for any number n, the bird $A\bar{n}$ is a propositional bird—one of the two birds t or f. A set S of numbers is said to be *computable* if there is a property bird A such that $A\bar{n}$ = t for every n in the set S, and $A\bar{n}$ = f for every n not in the set S. Such a bird A is said to *compute* the set S. And a set S is called *computable* if there is a bird A that computes it.

"The nice thing about a computable set S is that given any number n, you can find out whether n belongs to the set or whether it doesn't; you just go over to the bird A, which computes S, and call out $\bar{n}$. If A responds by calling out t, you know that n is in the set S; if A calls out f, you know that n is not in the set S.

"As an example, the set E of all even numbers is computable—there is a bird A such that $A\bar{0}$ = t; $A\bar{1}$ = f; $A\bar{2}$ = t; $A\bar{3}$ = f; and for *every* even number n, $A\bar{n}$ = t, whereas for *every* odd number n, $A\bar{n}$ = f. Can you see how to find A? You might try using the fixed point principle."

9 · THE BIRD G

"By a *relational* bird—or more properly, a relational bird of degree 2—is meant a bird A such that for any numbers a and b, $A\overline{a}\overline{b}$ = t or $A\overline{a}\overline{b}$ = f.

"You are probably familiar with the symbol >, meaning 'greater than,' " continued Griffin. "For any numbers a and b, we write a > b to mean that a is greater than b—so, for example, 8 > 5 is true; 4 > 9 is false; also 4 > 4 is false. We now need a relational bird that computes the relation 'is greater than'—that is, we need a bird g such that for any numbers a and b, if a > b, then gab = t, but if a $\leq$ b, meaning that a is less than or equal to b, then gab = f. Can you see how to find such a bird?

"This is a bit tricky," Griffin added, "so I had best point out the following facts. The relation a > b is the one and only relation satisfying the following conditions, for any numbers a and b:

1. If a = 0, then a > b is false.
2. If a $\neq$ 0, then:
   a. If b = 0, then a > b is true.
   b. If b $\neq$ 0, then a > b is true if and only if (a - 1) > (b - 1).

"Now, using the fixed point principle, do you see how to find the bird g?"

10 · THE MINIMIZATION PRINCIPLE

"Now comes an important principle known as the *minimization* principle," said Griffin.

"Suppose that A is a relational bird such that for every number n, there is at least one number m such that $A\overline{n}\overline{m}$ = t. Such a relational bird is sometimes called *regular*. If A is regular, then for every number n there is obviously the *smallest* number k such that $A\overline{n}\overline{k}$ = t. Well, the minimization principle is that given any regular relational bird A, there is a bird A′, called a *minimizer* of A, such that for every number n, A′$\overline{n}$ = k, when k is the *smallest* number such that $A\overline{n}\overline{k}$ = t. So, for example, if $A\overline{n}\overline{0}$ = f and $A\overline{n}\overline{1}$ = f and $A\overline{n}\overline{2}$ = f, but $A\overline{n}\overline{3}$ = t, then A′$\overline{n}$ = 3. Can you see how to prove the minimization principle?"

Craig thought about this for some time.

"I'd better give you some hints," said Griffin. "Given a regular bird A, first show how to find a bird A_1 such that for all numbers n and m, the following two conditions hold:

1. If $A\overline{n}\overline{m}$ = f, then $A_1\overline{n}\overline{m} = A_1\overline{n}\overline{m^+}$.
2. If $A\overline{n}\overline{m}$ = t, then $A_1\overline{n}\overline{m} = \overline{m}$.

"Then take A' to be $CA_1\overline{0}$, where C is the cardinal, and show that A' is a minimizer of A."

II · THE LENGTH MEASURER

"By the *length* of a number n," said Griffin, "we mean the number of digits in n, when n is written in ordinary base 10 notation. Thus the numbers from 0 to 9 have length 1; those from 11 to 99 have length 2; those from 100 to 999 have length 3, and so forth.

"We now need a bird ℓ that measures the length of any number—that is, we want ℓ to be such that for any number n, $\ell\overline{n} = \overline{k}$, when k is the length of n. So, for example, $\ell\overline{7}$ = 1; $\ell\overline{59}$ = 2; $\ell\overline{648}$ = 3. Can you see how to find the bird ℓ?"

Craig thought about this for some time. "Ah!" he finally said. "I get the idea! The length of a number n is simply the smallest number k such that $10^k > n$."

"Good!" said Griffin.

With this, the reader should have no trouble finding the bird ℓ.

12 · CONCATENATION TO THE BASE 10

"Now for the last problem of today," said Griffin. "For any numbers a and b, by a * b we mean the number which, when written in base 10 notation, consists of a in base 10 notation, followed by b in base 10 notation. For example, 53 * 796 = 53796; 280 * 31 = 28031."

"That's a curious operation on numbers!" said Craig.

"It is an important one, as you will see tomorrow," replied Griffin. "This operation is known as *concatenation to the base 10*. And now we need a bird ⊛ that computes this operation—that is, we want ⊛ to be such that for any numbers a and b, $\circledast\, \bar{a}\, \bar{b} = \overline{a * b}$. Do you see how to find such a bird?"

SOLUTIONS

1 • We first show that $\bar{0}$ is different from all the birds $\bar{1}, \bar{2}, \bar{3}, \ldots \overline{n^+}, \ldots$

Well, suppose there were some number n such that $\bar{0} = \overline{n^+}$. Then I = Vf$\bar{n}$. Then IK = Vf$\bar{n}$K = Kf$\bar{n}$ = f. Hence we would have K = KI, since IK = K and f = KI, but we already know that K $\neq$ KI. Therefore $\bar{0} \neq \overline{n^+}$.

We must next show that for any numbers n and m, if $\overline{m^+} = \overline{n^+}$, then m = n. Well, suppose that $\overline{n^+} = \overline{m^+}$. Then Vf$\bar{n}$ = Vf$\bar{m}$. Hence Vf$\bar{n}$f = Vf$\bar{m}$f, so ff$\bar{n}$ = ff$\bar{m}$, hence $\bar{n}$ = $\bar{m}$, since ff$\bar{n}$ = $\bar{n}$ and ff$\bar{m}$ = $\bar{m}$.

Now that we know that $\bar{0} \neq \overline{m^+}$ and that for every n and m, if $\overline{n^+} = \overline{m^+}$ then n = m, the proof that all the birds $\bar{0}, \bar{1}, \bar{2}, \ldots, \bar{n}, \ldots$ are distinct proceeds exactly as in the solution to Problem 19, Chapter 22.

2 • Take P to be Tf, where T is the thrush and f is the bird KI, as in the last chapter. Then for any number n, $P\overline{n^+}$ = Tf$\overline{n^+}$ = $\overline{n^+}$f = Vf$\bar{n}$f = ff$\bar{n}$ = $\bar{n}$.

3 • · Take Z to be Tt; T is the thrush, and t is the truth bird K. Then:

1. $Z\bar{0}$ = TtI = It = t. So $Z\bar{0}$ = t.

2. Now take any number n. Then $Z\overline{n^+}$ = Tt$\overline{n^+}$ = $\overline{n^+}$t = Vf$\bar{n}$t = tf$\bar{n}$ = f.

Note: Under the particular scheme used by Griffin for representing numbers by birds, the birds σ, P, and Z are relatively easy to find. This is the technical advantage to which Griffin referred. Any other scheme that would yield a successor bird, a predecessor bird, and a zero-tester would also work.

4 • The zero-tester Z is such a bird A! *Reason:* $Z\bar{0}$xy = txy, since $Z\bar{0}$ = t, and txy = x, so $Z\bar{0}$xy = x. But for any n $\neq$ 0, $Z\bar{n}$ = f, hence $Z\bar{n}$xy = fxy = y.

5 • The addition operation + is uniquely determined by the following two conditions, for any numbers n and m:

1. $n + 0 = n$
2. $n + m^+ = (n + m)^+$. That is, n plus the successor of m is the successor $n + m$.

We therefore seek a bird A such that for all n and m:

1. $A\bar{n}\bar{0} = \bar{n}$
2. $\overline{A\bar{n}m^+} = \sigma(\overline{A\bar{n}m})$, or what is the same thing, for any positive m, $A\bar{n}m = \sigma(A\bar{n}(P\bar{m}))$.

Thus A must satisfy the condition that for any n and any m, whether 0 or positive, $A\bar{n}\bar{m} = Z\bar{m}\bar{n}(\sigma(A\bar{n}(P\bar{m})))$. Such a bird A exists by the fixed point principle, and so we take $\oplus$ to be any such bird A.

6 • We note that multiplication is the one and only operation satisfying the following two conditions:

1. For any number n, $n \cdot 0 = 0$.
2. For any numbers n and m, $n \cdot m^+ = (n \cdot m) + n$. We therefore want a bird A such that for every n and m, $A\bar{n}\bar{m} = (Z\bar{m})\bar{0}((\oplus)(A(\bar{n}(P\bar{m}))\bar{n}))$. Again, such a bird A can be found by the fixed point principle and we take $\otimes$ to be such a bird.

7 • The exponential operation obeys the following well-known laws:

1. $n^0 = 1$
2. $n^{m^+} = n^m \times n$

We therefore seek a bird Ⓔ such that for all n, $Ⓔ\bar{n}\bar{0} = \bar{1}$, and for any positive number m, $Ⓔ\bar{n}\bar{m} = \otimes(Ⓔ\bar{n}\bar{m})\bar{n}$. Equivalently we want a bird Ⓔ such that for all n and m, $Ⓔ\bar{n}\bar{m} = Z\bar{m}\bar{1} \otimes (Ⓔ\bar{n}\bar{m})\bar{n})$. Again, such a bird Ⓔ can be found by the fixed point principle.

8 • The property of being an even number is the one and only property satisfying the following two conditions:

1. 0 is even.
2. For any positive number n, n is even if and only if its predecessor is *not* even.

We therefore seek a bird A such that:

1. $A\bar{0} = t$
2. For any positive n, $A\bar{n} = N(A(P\bar{n}))$, where N is the negation bird.

We thus want a bird A such that for every n, whether positive or 0, $A\bar{n} = Z\bar{n}t(N(A(P\bar{n})))$. Again such a bird A exists by the fixed point principle.

9 • By virtue of the conditions given, we seek a bird g such that for any numbers a and b:

1. If $Z\bar{a} = t$, then $g\overline{ab} = f$.
2. If $Z\bar{a} = f$, then:
   a. If $Z\bar{b} = t$, then $g\overline{ab} = t$.
   b. If $Z\bar{b} = f$, then $g\overline{ab} = g(P\bar{a})(P\bar{b})$.

Equivalently, we want a bird g such that for all numbers a and b, the following holds:

$$g\overline{ab} = Z\bar{a}f(Z\bar{b}t(g(P\bar{a})(P\bar{b})))$$

Again such a bird g exists by the fixed point principle.

10 • Suppose A is a regular relational bird. By the fixed point principle there is a bird A_1 such that for all birds x and y, $A_1xy = (Axy)y(A_1x(\sigma y))$. Then for any numbers n and m, $A_1\bar{n}\bar{m} = A(\bar{n}\bar{m})\bar{m}(A_1\bar{n}\bar{m}^+)$. Thus Condition 1 and Condition 2 hold, because the value of $A(\bar{n}\bar{m})\bar{m}(A_1\bar{n}\bar{m}^+)$ is $\bar{m}$, if $A\bar{n}\bar{m} = t$, and is $A_1\bar{n}\bar{m}^+$, if $A\bar{n}\bar{m} = f$.

Following Griffin's suggestion, we assume $A' = CA_1\bar{0}$. Then for every n, $A'\bar{n} = A_1\bar{n}\bar{0}$ (because $A'\bar{n} = CA_1\bar{0}\bar{n} = A_1\bar{n}\bar{0}$). Now, given an n, let k be the smallest number such that $A\bar{n}\bar{k} = t$. For example, suppose k = 3. Then $A\bar{n}\bar{0} = f$; $A\bar{n}\bar{1} = f$; $A\bar{n}\bar{2} = f$; but $A\bar{n}\bar{3} = t$. We must show that $A'\bar{n} = \bar{3}$—in other words, that $A_1\bar{n}\bar{0} = \bar{3}$. Well, since $A\bar{n}\bar{0} = f$, then $A_1\bar{n}\bar{0} = A_1\bar{n}\bar{1}$, by Condition 1. Since $A\bar{n}\bar{1} = f$, then $A_1\bar{n}\bar{1} = A_1\bar{n}\bar{2}$, again by Condition 1. Since $A\bar{n}\bar{2} = f$, then $A_1\bar{n}\bar{2} = A_1\bar{n}\bar{3}$, again by Condition 1. But now $A\bar{n}\bar{3} = t$, hence $A_1\bar{n}\bar{3} = \bar{3}$, by Condition 2. And so $A_1\bar{n}\bar{0} = A_1\bar{n}\bar{1} = A_1\bar{n}\bar{2} = A_1\bar{n}\bar{3} = \bar{3}$, therefore $A_1\bar{n}\bar{0} = \bar{3}$, and so $A'\bar{n} = \bar{3}$.

We illustrated the proof for k = 3, but the reader can readily see that the same type of proof would work if k were any other number.

11 • A single example should convince the reader of the correctness of Craig's assertions:

Suppose n = 647. The length of 647 is 3, and 10^3 = 1000, which is greater than 647. But 10^2 = 100, which is less than 647. Perhaps we should also consider the case when n itself is a power of 10—suppose n, say, is 100. Then 10^3 > 100, but 10^2, though not less than 100, is not greater than 100; it is equal to 100. So 3 is the smallest number such that 10^3 > 100.

Now to find the bird ℓ: We let A_1 be the bird $Bg(\text{\textcircled{E}}\overline{10})$, where B is the bluebird. Then for any numbers n and m, $Bg(\text{\textcircled{E}}\overline{10})\bar{n}\bar{m}$ = $g(\text{\textcircled{E}}\overline{10}\bar{n})\bar{m}$ = $g\overline{10}\bar{n}\bar{m}$, which is t if 10^n > m, and is f otherwise. And so A_1 is a relational bird such that $A_1\bar{n}\bar{m}$ = t if and only if 10^n > m. We then let A be the bird CA_1, where C is the cardinal. Then $A\bar{n}\bar{m}$ = $A_1\bar{n}\bar{m}$, and so $A\bar{n}\bar{m}$ = t if 10^m > n; otherwise $A\bar{n}\bar{m}$ = f. Finally, we take ℓ to be a minimizer of A, and so $\ell\bar{n}$ is the *smallest* m such that 10^m > n—in other words, $\ell\bar{n}$ = $\bar{k}$, where k is the length of n.

12 • We first illustrate the general idea with an example. Suppose a = 572 and b = 39. Then 572 * 39 = 57239 = 57200 + 39 = 572 · 10^2 + 39, and 2 is the length of 39.

In general, a * b = a · 10^k + b, where k is the length of b. We accordingly take ⊛ to be a bird such that for all x and y, ⊛xy = $\oplus(\otimes x(\text{\textcircled{E}}\overline{10}(\ell y)))y$. As the reader can easily verify, for any numbers a and b, $⊛\bar{a}\bar{b}$ = a · 10^k + b, where k is the length of b.

CHAPTER 25

IS THERE AN IDEAL BIRD?

"Tomorrow I unfortunately must leave," said Craig, "but before I do, I want to tell you of a problem I have been unable to solve. Perhaps you may know the answer.

"Any expression X, built from the symbols S and K, and parenthesized correctly, is the name of some bird. Now, two different expressions might happen to name the same bird—for example, the expressions ((SK)K)K and KK(KK) are both names of the kestrel K, though the expressions themselves are different. Now, what I want to know is this: Given two expressions X_1 and X_2, is there any systematic way of determining whether or not they name the same bird?"

"A beautiful question!" replied Griffin. "And it's an amazing coincidence that you ask it today. This was just the topic I was planning to talk to you about. This question has been on the minds of some of the world's ablest logicians and has come to be known as the *Grand Question*.

"To begin with, any question as to whether two expressions name the same bird can be translated into a question of whether a certain number belongs to a certain set of numbers."

"How is that?" asked Craig.

"This is done using a clever device attributable to Kurt Gödel—the device known as *Gödel numbering*, which I will shortly explain."

"All the birds here are derivable from S and K, and their behavior—the way a bird x responds to a bird y—is strictly determined by the rules of combinatory logic. Combinatory logic is a theory that can be completely formalized. The theory uses just five symbols:

$$S \quad K \quad (\quad) \quad =$$
$$1 \quad 2 \quad 3 \quad 4 \quad 5$$

"Under each symbol I have written the number called its *Gödel* number, but I'll tell you more about Gödel numbering a bit later.

"Any expression built from the two letters S and K and parenthesized correctly is called a *term*. To be more precise, a term is any expression in the first four symbols that is constructed by the following two rules:

1. The letters S and K standing alone are terms.

2. Given any terms X and Y already constructed, we may form the new term (XY).

"In application to this bird forest, the *terms* are those expressions that are names of birds. The letter S is the name of one particular starling—which one doesn't really matter—and the letter K is the name of one particular kestrel.

"By a *sentence* is meant an expression of the form X = Y, when X and Y are terms. The sentence X = Y is called *true* if X and Y are names of the same bird *and false* otherwise. In order for a sentence X = Y to be true, the term X doesn't have to be the same as the term Y; it merely suffices that these terms name the same bird.

"Of course, for any terms X, Y, and Z, the sentence SXYZ = XZ(YZ) is true, by definition of the starling, and KXY = X is true, by definition of the kestrel. All such sentences are taken as *axioms* of combinatory logic. We also take as axioms all sentences of the form X = X; these sentences are trivially true. These are the only axioms we shall take. We then *prove* various sentences to be true by starting with the axioms and using the usual logical rules for equality, which are:

1. If we can prove X = Y, then we can conclude that Y = X.

2. If we can prove X = Y and Y = Z, then we can conclude that X = Z.

3. If we can prove X = Y, then for any term Z we can conclude that XZ = YZ and that ZX = YX.

"Now, when I said that the behavior of the birds of this forest is completely determined by the laws of combinatory logic, what I meant is that a sentence X = Y is true, in the sense that the terms X and Y name the same bird, if and only if the sentence X = Y is *provable* from the above axioms by the rules I have just mentioned. There are no 'accidental' relations between our birds; X = Y only if the fact is *provable*.

"This system of combinatory logic is known to be consistent, in the sense that not every sentence is provable—in particular, the sentence KI = K is not provable. If this one sentence were provable, then every sentence would be provable by essentially the same argument we used to show that if KI = K, then there could be only one bird in the forest. We will use f to abbreviate KI, and we will also use t synonymously with K, and so the sentence f = t is an important example of a sentence that is not provable in the system.

"And now for Gödel numbering: I have already told you that the Gödel numbers of the five symbols S, K, (,), and = are respectively 1, 2, 3, 4, and 5. The Gödel number of any compound expression is obtained by simply replacing each symbol with the digit representing its Gödel number and then reading off the resulting string of digits to the base 10. For example, the expression (SK) consists of the third symbol, followed by the first symbol, followed by the second symbol, followed by the fourth symbol, and so its Gödel number is 3124—three thousand one hundred twenty-four.

"Now, let $\mathcal{T}$ be the set of Gödel numbers of the true sentences. Given any terms X and Y, they name the same bird if and only if the sentence X = Y is true, and the sentence is true if and only if its Gödel number lies in the set $\mathcal{T}$. That's what I meant when I said that any question of whether or not two terms X and Y name the same bird can be translated into a question of whether a certain number—namely, the Gödel number of the sentence X = Y—lies in a certain set of numbers—namely, the set $\mathcal{T}$.

"Now, the question *you* are asking boils down to this: Is the set $\mathscr{T}$ a computable set? Is there some purely deterministic device that can compute which numbers are in $\mathscr{T}$ and which ones are not? As I have told you, anything a computer can do can be done by one of our birds, and so your question is equivalent to this: Is there here some 'ideal' bird A that can evaluate the truth of all sentences of combinatory logic? Is there a bird A such that whenever you call out the Gödel number of a true sentence, the bird will call back "t," and whenever you call out any other number, the bird will call back "f"? In other words, is there a bird A such that for every n in $\mathscr{T}$ A$\bar{n}$ = t and for every n not in $\mathscr{T}$, A$\bar{n}$ = f? That is the question you are asking. Such a bird could settle *all* formal mathematical questions, because all such questions can be reduced to questions of which sentences are provable in combinatory logic and which sentences are not. Combinatory logic is a *universal* system for all of formal mathematics, and so any ideal bird might be said to be mathematically omniscient. That is why so many people have come to this forest in search of this bird."

"That is staggering to the imagination!" said Craig. "Is it yet known whether or not there is this 'ideal' bird?"

"The question, in one form or another, has been on the minds of many mathematicians and philosophers from Leibniz on—and possibly earlier. It can be equivalently formulated: Can there be a *universal* computer that can settle all mathematical questions? Thanks to the work of Gödel, Church, Turing, Post, and others, the answer to this question is now known once and for all. I won't spoil it for you by telling you the answer yet, but before this day is over, you will know the answer.

"We did a good deal of the preliminary work yesterday when we derived the concatenation bird ⊛, but there are still a few preliminaries left before we can answer the Grand Question.

"You realize, of course, that for any expressions X and Y, if a is the Gödel number of X and b is the Gödel number of Y, then the Gödel number of XY is a * b. For example, suppose X is the expression S and Y is the expression K. The Gödel number of X is 31 and the Gödel number of Y is 24. The expression XY is (SK) and its Gödel number is 3124, which is 31 * 24. Now you can see the significance of the numerical operation of concatenation to the base 10."

1 · NUMERALS

"By a *numeral* is meant any of the terms $\bar{0}, \bar{1}, \bar{2}, \ldots, \bar{n}, \ldots$ We call $\bar{n}$ the *numeral* for the number n. The term $\bar{n}$, like any other term, has a Gödel number; we let $n^{\#}$ be the Gödel number of the numeral $\bar{n}$.

"For example, $\bar{0}$ is I, which in terms of S and K is the expression $((SK)K)$; this expression has Gödel number 3312424. And so $0^{\#} = 3312424$.

"As for $1^{\#}$, this is already quite a large number: $\bar{1}$ is the expression $\sigma\bar{0}$, where σ is the expression (Vf), which in terms of S and K can be seen to be the expression $(S(K(S(S(S((SK)K)(K(K((SK)K)))))))K)$—a horrible expression whose Gödel number is 3132313133124243232331242444444424. To avoid having to write this number again, I will henceforth represent it by the letter s. Thus s is the Gödel number of σ. Then, since $\bar{1}$ is the expression $(\sigma\bar{0})$, the number $1^{\#}$ is $3 * s * 0^{\#} * 4$. Then $2^{\#} = 3 * s * 1^{\#} * 4$; $3^{\#} = 3 * s * 2^{\#} * 4$, and so forth. For each number n, $(n + 1)^{\#} = 3 * s * n^{\#} * 4$.

"What we now need is a bird such that when any number n is called to the bird, the bird will call back the number $n^{\#}$. That is, we want a bird δ such that for every number n, $\delta\bar{n} = \overline{n^{\#}}$. Can you see how to find such a bird δ?"

2 · NORMALIZATION

"For any expression X," said Griffin, "by $\ulcorner X \urcorner$ is meant the *numeral* designating the Gödel number of X. Thus $\ulcorner X \urcorner$ is $\bar{n}$, where n is the Gödel number of X. We call $\ulcorner X \urcorner$ the Gödel *numeral* of X.

"By the *norm* of X is meant the expression $X \ulcorner X \urcorner$—that is, X followed by its own Gödel numeral. If n is the Gödel number of X, then $n^{\#}$ is the Gödel number of $\ulcorner X \urcorner$ and so $n * n^{\#}$ is the Gödel number of $X \ulcorner X \urcorner$—the norm of X. And so if X has Gödel number n, then the norm of X has Gödel number $n * n^{\#}$.

"We now need a bird Δ called a *normalizer* such that for every number n, $\Delta n = n * n^{\#}$. This bird is easy to find, now that we have the birds $\circledast$ and δ. Can you see how?"

3 · THE SECOND FIXED POINT PRINCIPLE

"One can do some amazing things with the normalizer," said Griffin. "I will give you an example.

"We shall say that a term X *designates* a number n if the sentence X = $\bar{n}$ is true. Obviously, one term that designates n is the numeral $\bar{n}$, but there are infinitely many others. For example, I$\bar{n}$, I(I$\bar{n}$), I(I(I$\bar{n}$)), . . . are all terms that designate n. Also, taking 8 for n, the numeral 8 designates 8; so does the term $\oplus\bar{2}\bar{6}$; so does the term $\oplus\bar{3}\bar{5}$; so does the term $\otimes\bar{2}\bar{4}$. You get the idea!

"We call a term a *numerical* term if it designates some number n. Every numeral is a numerical term, but not every numerical term is a numeral— for example, the expression $\oplus\bar{2}\bar{6}$ is a numerical term, but it is not a numeral. For any number n, there is only one numeral designating it—the numeral $\bar{n}$—but there are infinitely many numerical terms designating it.

"Now, it is impossible that any numeral can designate its own Gödel number, because for any number n, the Gödel number of the numeral $\bar{n}$ is much larger than n. All I am saying is that for every n, $n^{\#} > n$. However, there *does* exist a *numerical term* X that designates its own Gödel number."

"That's surprising!" said Craig. "I have no idea why this should be."

"One can also construct a term that designates twice its Gödel number," said Griffin, "or one that designates three times its Gödel number, or one that designates five times its Gödel number plus seven. All these odd facts are special cases of a very important principle known as the *second fixed point principle,* which is this: For any term A, there is a term X such that the sentence A$\ulcorner$X$\urcorner$ = X is true. Stated otherwise, for any term A there is a term X such that X names the same bird as A followed by the Gödel numeral of X.

"Can you see how to prove this? Also, can you see how the oddities I just mentioned are special cases of the second fixed point principle?"

4 · A Gödelian Principle

"The second fixed point principle yields as a corollary an important principle attributable to Gödel, which I will tell you in a moment," said Griffin.

"For any set $\mathscr{S}$ of numbers, a sentence X is called a *Gödel sentence* for $\mathscr{S}$ if either X is true and its Gödel number is in $\mathscr{S}$, or X is false and its Gödel number is not in $\mathscr{S}$. Such a sentence can be thought of as expressing the proposition that its own Gödel number is in $\mathscr{S}$, because the sentence is true if and only if its Gödel number *is* in $\mathscr{S}$.

"Now, Gödel's principle is this: For any computable set $\mathscr{S}$, there is a Gödel sentence for $\mathscr{S}$. For example, since the set of even numbers is computable, then there must be a sentence such that either it is true and its Gödel number is even, or it is false and its Gödel number is odd. Again, since the set of odd numbers is computable, then there must be a sentence such that either it is true and its Gödel number is odd, or it is false and its Gödel number is even. The remarkable thing is that for *any* computable set, there is a Gödel sentence for that set. This follows fairly easily from the second fixed point principle. Do you see how?

"I'll give you a hint," added Griffin. "For any set $\mathscr{S}$, let $\mathscr{S}^*$ be the set of all numbers n such that n ∗ 52 is in $\mathscr{S}$. First prove as a lemma—a preliminary fact—that if *if* is computable, so is $\mathscr{S}^*$."

"What is the significance of the number n ∗ 52?" asked Craig.

"If n is the Gödel number of an expression X," replied Griffin, "then n ∗ 52 is the Gödel number of the expression X = t."

How is Gödel's principle proved?

5 · The Negation Bird Pops Up

"One last detail before we answer the Grand Question," said Griffin. "For any set $\mathscr{S}$ of numbers, by $\mathscr{S}'$ we mean the set of all numbers not in $\mathscr{S}$. For example, if $\mathscr{S}$ is the set of all even numbers, $\mathscr{S}'$ is the set of all odd numbers. The set $\mathscr{S}'$ is called the *complement* of $\mathscr{S}$.

"Prove that if $\mathscr{S}$ is computable, so is $\mathscr{S}'$."

· 6 ·

"Now we have all the pieces of the puzzle," said Griffin. "We are letting $\mathscr{T}$ be the set of Gödel numbers of all the true sentences. First ask yourself whether there could possibly be any Gödel sentence for the complement $\mathscr{T}'$ of $\mathscr{T}$. Then, using the last two results, show that the set $\mathscr{T}$ is *not* computable."

"This is amazing indeed!" said Craig, after he realized the solution. "It seems to shatter any hope of a purely mechanical device that can decide all mathematical questions."

"It certainly does!" said Griffin. "Any such mechanism could determine which numbers are in $\mathscr{T}$ and which ones are not, hence $\mathscr{T}$ would be a computable set, which we have just seen is not the case. Since $\mathscr{T}$ is not computable, then there is no mechanism that can compute it. In short, no mechanism can be mathematically omniscient.

"Since $\mathscr{T}$ is not computable," continued Griffin, "then no bird of this forest can compute it, and so there is no ideal bird here. Despite the cleverness of many of our birds, none of them is mathematically omniscient.

"But you know," said Griffin, with a dreamy look in his eyes, "there has been a rumor that in the days before I came here, a bird from another forest far, far away once visited this forest and astounded all the other birds by appearing to be mathematically omniscient. Of course, this is only a rumor, but who really knows? If the rumor is true, then that bird must have been most remarkable; no purely mechanistic explanation could possibly account for its behavior. Those philosophers who are mechanistically oriented and believe that birds, humans, and all other biological organisms are nothing more than elaborate mechanisms would of course deny that any such bird is possible. But I, who do not have complete confidence in the philosophy of mechanisms, reserve judgment on the matter. I'm not saying that I believe the rumor; I'm not saying that there is or was such a bird; I'm merely saying that I believe such a bird might be possible.

"I wish we had more time," concluded Griffin. "There are so many more facts about this forest that I believe would interest you."

"I have no doubt!" said Craig, rising. "I am infinitely grateful for all you have taught me, and I'm hoping that I might be able to visit this forest again one day."

"That would be wonderful!" said Griffin.

Craig left the forest the next day with a twinge of sadness. Although part of him looked forward to renewing his more normal life of crime detection, Craig realized that in his advancing years his interests were veering more and more to the purely abstract and theoretical.

"This vacation has been like an idyllic dream," thought Craig, as he reached the exit—also the entrance—gate. "I really *must* visit this forest again!"

"Only the elite are allowed to leave this forest!" said an enormous sentinel who blocked his way. "However, since you have entered this forest and only the elite are allowed to enter, then you must be one of the elite. Therefore you are free to leave, and God speed you well!"

"This is one ritual I will never understand," thought Craig, as he shook his head with an amused smile.

Solutions

1 • We first need a bird A such that for any number n, $A\bar{n} = \overline{3 * s * n * 4}$. We can take A to be $B(C \circledast \bar{4})(\circledast \overline{3 * s})$, where B is the bluebird and C is the cardinal.

Now we want a bird δ such that for every n, if n = 0 then $\delta\bar{n} = \overline{0^{\#}}$, and if n >0, then $\delta\bar{n} = A(P\bar{n})$. Equivalently, we want a bird δ such that for all n, $\delta\bar{n} = (Z\bar{n})\overline{0^{\#}}(A(\delta(P\bar{n})))$. Such a bird δ can be found by the fixed point principle.

2 • We take Δ to be the bird $W(DC \circledast \delta)$, where W is the warbler, D is the dove, and C is the cardinal. Then for any number n, $\Delta\bar{n} = W(DC \circledast \delta)\bar{n}$ $= DC \circledast \delta\bar{n}\bar{n} = C \circledast (\delta\bar{n})\bar{n} = \circledast\bar{n}(\delta\bar{n}) = \circledast\bar{n}\overline{n^{\#}} = \overline{n * n^{\#}}$.

3 • Let Δ be the normalizing bird—or, more precisely, the term $W(DC \circledast \delta)$, which names the normalizing bird. Then for any expression X, the sentence $\Delta \ulcorner X \urcorner = \ulcorner X \ulcorner X \urcorner \urcorner$ is true, because X has some Gödel number n; $X \ulcorner X \urcorner$ has Gödel number $n * n^{\#}$, so the above sentence is $\Delta \bar{n} = \overline{n * n^{\#}}$.

Now take any term A. Let X be the term $BA\Delta \ulcorner BA\Delta \urcorner$, where B is a term for the bluebird. We now show that the sentence $A \ulcorner X \urcorner = X$ is true.

The sentence $BA\Delta \ulcorner BA\Delta \urcorner = A(A \ulcorner BA\Delta \urcorner)$ is obviously true. Also the sentence $\Delta \ulcorner BA\Delta \urcorner = \ulcorner BA\Delta \ulcorner BA\Delta \urcorner \urcorner$ is true, hence the sentence $A(A \ulcorner BA\Delta \urcorner) = A \ulcorner BA\Delta \ulcorner BA\Delta \urcorner \urcorner$ is true, and so the sentence $BA\Delta \ulcorner BA\Delta \urcorner = A \ulcorner BA\Delta \ulcorner BA\Delta \urcorner \urcorner$ is true. This is the sentence $X = A \ulcorner X \urcorner$, and so the sentence $A \ulcorner X \urcorner = X$ is true. This proves the second fixed point principle.

As an application, let us take I for A. Then there is a term X such that $I \ulcorner X \urcorner = X$ is true, hence $\ulcorner X \urcorner = X$ is true, and hence the sentence $X = \ulcorner X \urcorner$ is true. If we let n be the Gödel number of X, then the sentence $X = \bar{n}$ is true, and so X designates its own Gödel number n. By the above proof, we can take X to be the term $BI\Delta \ulcorner BI\Delta \urcorner$. However, there is a simpler term that designates its Gödel number—namely, $\Delta \ulcorner \Delta \urcorner$.

A term that designates twice its own Gödel number is $B(\otimes \bar{2})\Delta \ulcorner B(\otimes \bar{2})\Delta \urcorner$. Why?

4 • Let us first prove the lemma. For any bird A, let $A^{\#}$ be the bird $BA(C \circledast \overline{52})$, where B is the bluebird and C is the cardinal. For any number n, $A^{\#}n = A\overline{n * 52}$, because $A^{\#}n = BA(C \circledast \overline{52})\bar{n} = A(C \circledast \overline{52}\bar{n}) = A(\circledast \bar{n}\overline{52}) = A\overline{n * 52}$. This proves that $A^{\#}\bar{n} = A\overline{n * 52}$.

Now suppose A computes $\mathscr{S}$. Then $A^{\#}$ must compute S^*, because for any n in $\mathscr{S}^*$, the number $n * 52$ is in $\mathscr{S}$, hence $A\overline{n * 52} = t$, and so $A^{\#}\bar{n} = t$. Also, for any number n not in $\mathscr{S}^*$, the number $n * 52$ is not in $\mathscr{S}$, hence $A\overline{n * 52} = f$, and so $A^{\#}\bar{n} = f$. This proves that $A^{\#}$ computes $\mathscr{S}^*$.

Now for the proof of Gödel's principle. Suppose $\mathscr{S}$ is computable. Then $\mathscr{S}^*$ is computable, as we have just seen. Let A be a bird that computes $\mathscr{S}^*$. By the second fixed point principle there is a term X such that the sentence $A^{\ulcorner}X^{\urcorner} = X$ is true. Let Y be the sentence $X = t$. We will show that Y is a Gödel sentence for the set $\mathscr{S}$.

Let n be the Gödel number of X. Then Y, being the sentence $X = t$, has Gödel number $n * 52$.

a. Suppose that Y is true. Then the sentence $X = t$ is true and since the sentence $A^{\ulcorner}X^{\urcorner} = X$ is also true, then the sentence $A^{\ulcorner}X^{\urcorner} = t$ is true, and so the sentence $A\bar{n} = t$ is true (because $^{\ulcorner}X^{\urcorner}$ is the numeral $\bar{n}$). Therefore n belongs to the set $\mathscr{S}^*$ (because A computes $\mathscr{S}^*$, hence if n didn't belong to $\mathscr{S}^*$, then the sentence $A\bar{n} = f$ would be true, which is impossible, since $A\bar{n} = t$ is true). Since n belongs to $\mathscr{S}^*$, then $n * 52$ belongs to $\mathscr{S}$, but $n * 52$ is the Gödel number of the sentence Y! This proves that if Y is true, then its Gödel number $n * 52$ belongs to $\mathscr{S}$.

b. Conversely, suppose $n * 52$ belongs to $\mathscr{S}$. Then n belongs to $\mathscr{S}^*$, hence $A\bar{n} = t$ is true, which means that Y is true. And so if the Gödel number of Y is in $\mathscr{S}$, then Y is true, or what is the same thing, if Y is false, then its Gödel number does not belong to $\mathscr{S}$.

According to argument a and argument b, we see that if Y is true, then its Gödel number is in $\mathscr{S}$ and if Y is false, then its Gödel number is not in $\mathscr{S}$. And so Y is a Gödel sentence for $\mathscr{S}$.

5 • Let A compute $\mathscr{S}$. Then BNA computes $\mathscr{S}'$, where B is the bluebird and N is the negation bird. *Reason:* For any number n, $BNA\bar{n} = N(A\bar{n})$. If n belongs to $\mathscr{S}'$ then n doesn't belong to $\mathscr{S}$, hence $A\bar{n} = f$, hence $N(A\bar{n}) = t$, so $BNA\bar{n} = t$. If n doesn't belong to $\mathscr{S}'$, then n belongs to $\mathscr{S}$, hence $A\bar{n} = t$, hence $N(A\bar{n}) = f$, and so $BNA\bar{n} = f$. Therefore BNA computes $\mathscr{S}'$.

6 • There certainly cannot be any Gödel sentence Y for the set $\mathscr{T}'$, because if Y is true, then its Gödel number is in $\mathscr{T}$, not in $\mathscr{T}'$, and if Y is false, its Gödel number is in $\mathscr{T}'$, not in $\mathscr{T}$. Therefore there is no Gödel sentence for $\mathscr{T}'$.

Now, if $\mathscr{T}$ were computable, then $\mathscr{T}'$ would be computable by Problem 5, hence by Problem 4 there would be a Gödel sentence for $\mathscr{T}'$. Since there is no Gödel sentence for $\mathscr{T}'$, then the set $\mathscr{T}$ is not computable.

CHAPTER 26

EPILOGUE

Inspector Craig arrived home not long afterward, and the first thing he did (after solving the case of the bat and the Norwegian maid) was to spend a long holiday weekend with his old friends McCulloch and the logician Fergusson.[*] He told them the entire story of his summer adventures.

"I have known nothing about combinatory logic till now," said McCulloch, "and I must say that the subject intrigues me enormously. But I would like to know how, when, and why the field ever got started. What was the motivation, and are there any practical applications?"

"Many," replied Fergusson (who was quite knowledgeable about all this). "Why, these days combinatory logic is one of the big things in computer science and artificial intelligence. The study of combinators started early in the twenties, pioneered by Shönfinkel. It is curious that *schön* in German means "beautiful," and *finkel* means "bird," hence *Shönfinkel* means "beautiful bird." So perhaps there's been a connection between birds and combinators all along! At any rate, the subject was further developed by Curry, Fitch, Church, Kleene, Rosser, and Turing, and in later years by Scott, Seldin, Hindley, Barandregt, and others. Their interests were purely theoretical; they were exploring the innermost depths

---

[*]A complete account of McCulloch's remarkable number machines and Fergusson's logic machines can be found in *The Lady or the Tiger?* (Alfred A. Knopf, 1982).

of logic and mathematics. No one then could have dreamed of the impact the subject would one day have on computer science. In recent times, the subject has been put on a more solid foundation—largely through the efforts of the logician Dana Scott, who provided interesting models for the theory."

"How is combinatory logic related to computer science?" asked Craig. "Professor Griffin didn't say too much about that."

"Why, in the construction of *programs,*" replied Fergusson. "Computers run on programs, you know, and these days all computer programs can be written in terms of combinators. The essential idea is that, given any programs X and Y, we can obtain a new program by feeding Y as input to the computer whose program is X; the resulting output is the program XY. The situation is analogous to calling out the name of one of Griffin's birds y to a bird x and getting the name of the bird xy as a response. The analogy is quite exact: Just as all combinatorial birds are derivable from the two birds S and K, so are *all* computer programs expressible in terms of the basic combinators S and K. We have here a case of what mathematicians call *isomorphism,* which in this instance means that the birds of Griffin's forest can be put into a one-to-one correspondence with all computer programs in such a manner that if a bird x corresponds to a program X and a bird y corresponds to a program Y, then the bird xy will correspond to the program XY. This is what Griffin must have meant when he said that, given any computer, there is a bird in his forest which can match it.

"I can certainly see," concluded Fergusson, "why Griffin has no need of computers: Because of the isomorphism of his forest of birds to the class of computer programs, it follows that any information a computer scientist can obtain from running his programs, Griffin can get just as surely by interrogating his birds. And yet Griffin's ideals seem to contrast strangely with those of people working in artificial intelligence. The latter are trying to simulate the thinking of biological organisms. Griffin is now turning the tables by using biological organisms—birds, in this case—to do the work of clever mechanisms. I believe the two approaches cannot but supplement each other, and it should be extremely interesting to see the outcome of all this!"

Many years later, Craig did indeed return to the Master Forest. But that is another story.

WHO'S WHO AMONG THE BIRDS

Bluebird	$Bxyz = x(yz)$
Cardinal	$Cxyz = xzy$
Dove	$Dxyzw = xy(zw)$
Eagle	$Exyzwv = xy(zwv)$
Finch	$Fxyz = zyx$
Goldfinch	$Gxyzw = xw(yz)$
Hummingbird	$Hxyz = xyzy$
Identity bird	$Ix = x$
Jay	$Jxyzw = xy(xwz)$
Kestrel	$Kxy = x$
Lark	$Lxy = x(yy)$
Mockingbird	$Mx = xx$
Owl	$Oxy = y(xy)$
Queer bird	$Qxyz = y(xz)$
Quixotic bird	$Q_ixyz = x(zy)$
Quirky bird	$Q_3xyz = z(xy)$
Robin	$Rxyz = yzx$
Sage bird	$\Theta x = x(\Theta x)$
Starling	$Sxyz = xz(yz)$
Thrush	$Txy = yx$
Turing bird	$Uxy = y(xxy)$
Vireo	$Vxyz = zxy$
Warbler	$Wxy = xyy$
Converse warbler	$W'xy = yxx$

231

STARRED BIRDS

Cardinal once removed	$C^*xyzw = xywz$
Cardinal twice removed	$C^{**}xyzwv = xyzvw$
Warbler once removed	$W^*xyz = xyzz$
Warbler twice removed	$W^{**}xyzw = xyzww$

DERIVATIONS OF CERTAIN BIRDS FROM OTHERS

From B

Dove	BB
Eagle	B(BBB)

From B and T

Robin	BBT
Cardinal	RRR—also B(T(BBT))(BBT).
Finch	ETTET—also B(TT)(B(BBB)T).
Vireo	BCT—also CF.
Queer bird	CB
Quixotic bird	BCB
Quirky bird	BT
Goldfinch	BBC

From B, T, and M

Double mockingbird	BM
Lark	QM
Warbler	C(BMR)
Converse warbler	BMR—also CW.
Hummingbird	BW(BC)
Starling	B(BW)(BBC)—also BW*G.
Owl	QQW—also BWQ and SI.
Turing bird	LO—also L(SI).

Starred Birds

C*	BC
C**	BC*
W*	BW
W**	BW*

Some Sage Birds

BML	LO(LO)	BM(BWM)
Q(QM)M	W(QL(QL))	BM(RMB)
SLL	W(M(QL))	BM(CBM)
UU	WS(BWB)	

BOOKS BY RAYMOND SMULLYAN

Theory of Formal Systems (1961)

First-Order Logic (1968)

The Tao is Silent (1977)

What Is the Name of This Book? The Riddle of Dracula and Other Logical
  Puzzles (1978)

The Chess Mysteries of Sherlock Holmes (1979)

This Book Needs No Title (1980)

The Chess Mysteries of the Arabian Knights (1981)

Alice in Puzzle-Land (1982)

The Lady or the Tiger? (1982)

5000 B.C. and Other Philosophical Fantasies (1983)

To Mock a Mockingbird (1985)

Forever Undecided (1987)

Gödel's Incompleteness Theorems (1992)

Satan, Cantor and Infinity (1992)

Recursion Theory for Metamathematics (1993)

Diagonalization and Self-Reference (1994)

Set Theory and the Continuum Problem (1996)

The Riddle of Scheherazade (1997)

Some Interesting Memories: A Paradoxical Life (2002)

Who Knows?: A Study of Religious Consciousness (2003)

In Their Own Words: Pianists of the Piano Society (with Peter Bispham) (2009)

Logical Labyrinths (2009)

Rambles Through My Library (2009)

A Spiritual Journey: Reflections on the Philosophy of Religion, A Transcendental Journey,
  and Cosmic Consciousness Redux (2009)

King Arthur in Search of his Dog (2010)

The Godelian Puzzle Book: Puzzles, Paradoxes and Proofs (2013)

A Beginner's Guide to Mathematical Logic (2014)

The Magic Garden of George B. And Other Logic Puzzles (2015)

Reflections: The Magic, Music and Mathematics of Raymond Smullyan (2015)

A Beginner's Further Guide to Mathematical Logic (2016)

A Mixed Bag: Jokes, Puzzles, Riddles, & Memorabilia (2016)

Raymond Smullyan (1919–2017) was a mathematician, concert pianist, magician, and author of numerous books of logic puzzles, chess puzzles, mathematics, philosophy and memoir. His startlingly brilliant and original puzzle books, of a genre largely invented by him, won him a global following. They are both delightfully approachable and devilishly challenging, leading the reader by degrees from elementary puzzles to deep results in mathematical logic.